T0212762

SpringerBriefs in Electrical and Computer Engineering

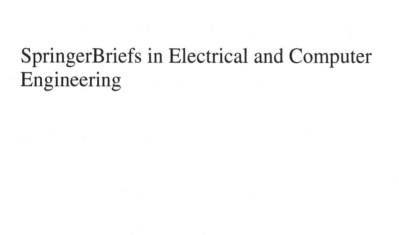

More information about this series at http://www.springer.com/series/10059

Yun Liao • Lingyang Song • Zhu Han

Listen and Talk

Full-duplex Cognitive Radio Networks

 Springer

Yun Liao
Peking University
Beijing, China

Lingyang Song
Peking University
Beijing, China

Zhu Han
Department of Engineering
University of Houston
Houston, TX, USA

ISSN 2191-8112   ISSN 2191-8120   (electronic)
SpringerBriefs in Electrical and Computer Engineering
ISBN 978-3-319-33977-1   ISBN 978-3-319-33979-5   (eBook)
DOI 10.1007/978-3-319-33979-5

Library of Congress Control Number: 2016940840

Printed on acid-free paper

This Springer imprint is published by Springer Nature
The registered company is Springer International Publishing AG Switzerland

Preface

With the proliferation of wireless services and the ever-increasing demand of data rate, spectrum scarcity has become a significant impediment to the development of diverse communication technology and applications. On the other hand, however, as suggested by technical reports from the Federal Communications Commission (FCC) and some other companies, many allocated spectrum bands are largely under-utilized in vast temporal and geographic dimensions. This is a direct result of the current spectrum allocation policy, which grants fixed spectrum bands to licensed users for exclusive use, leaving little space for any flexible adaptations.

The concept of cognitive radio that a device is able to observe the nearby wireless environment, and intelligently adapts its transmit/receive parameters for optimal spectrum usage, has attracted wide attention since its emergence. In a cognitive radio network (CRN), cognitive users (secondary users, SUs) are allowed to access spectrum bands of the licensed legitimate users (primary users, PUs) as long as they do not cause harmful interference at the PUs. Underlay and overlay are two types of access strategies for SUs. By underlay access, SUs can transmit concurrently with the PUs, but they need to adjust their transmit parameters to guarantee that the interference at the PU can be regarded as noise. By overlay access, SUs can only utilized the unoccupied spectrum holes, but the transmit rate is unconstrained. Besides, reliable identification of the spectrum holes is an essential overlay access.

Most existing work about overlay cognitive radio employ a "listen-before-talk" (LBT) protocol half-duplex (HD) radio. That means the SUs' traffic is time-slotted. The SUs listen to the bands for potential PUs' signal at the beginning of each slot, and then either access or back off according to the sensing results. This LBT protocol actually dissipates the precious resources by employing time-division duplexing and, thus, unavoidably suffers from two major problems: (1) transmission time reduction due to sensing and (2) sensing accuracy impairment due to data transmission.

In this book, we seek to bypass the above two problems by introducing full-duplex (FD) technology into CRNs. By FD technology, a wireless device can send and receive signals at the same time and frequency resources. Note that this was commonly considered impractical in the past due to the strong signal leakage from local output to input until the progress of self-interference suppression techniques recently. With FD radios, the SUs are able to sense the spectrum bands when they are transmitting, i.e., simultaneously listen and talk, showing the potential to significantly improve the spectrum utilization. To fully enjoy the benefits brought by FD technology, proper design of PHY and MAC layer protocols is highly needed, by which the attractive features of FD communications can be fully exploited, while the impact of the drawbacks like self-interference can be alleviated. In the protocol design, the secondary network can operate in various modes. For example, regarding the SUs' behavior, the FD SUs can either cooperate or compete with each other for spectrum access; and regarding the network architecture, there may or may not exist some secondary base stations in charge of all SUs. Moreover, if we go one step further, consider the case where the primary network keeps shutdown, the FD CRNs can be transformed into an FD WiFi network, which can also be an interesting topic that needs investigation.

The main aim of this book is to present a systematic picture of the FD CRNs and some related extensions. Specifically, we elaborate a "listen-and-talk" (LAT) protocol designed for PHY layer of FD CRNs in Chap. 2 and report a trade-off between secondary transmit power and throughput due to the existence of residual self-interference. Then in Chap. 3, we manage to extend the basic LAT protocol into multiuser scenarios and discuss the cooperation and contention mechanisms between FD SUs. In Chap. 4, we study a further extension of FD WiFi networks. By directly applying the LAT protocol to it, we investigate the PHY layer sensing performance. Besides, a feasible MAC layer protocol is designed, and the analysis shows that the spectrum can be almost fully utilized if the transmit parameters is carefully selected by using the proposed MAC protocol. Finally, in Chap. 5, apart from drawing a conclusion to the entire book, we list some challenging problems such as FD MIMO networks and the coexistence issue of FD cognitive networks and conventional networks, all of which wait to be investigated in the future.

Beijing, China   Yun Liao
Beijing, China   Lingyang Song
Houston, TX, USA   Zhu Han

Contents

Chapter 1
Introduction

1.1 Overview of Full-Duplex Radio

The wireless revolution has resulted in ever-increasing demands on the limited wireless spectrum, driving the request for systems with higher spectral efficiency. Among the emerging technologies for next-generation wireless networks, *in-band full-duplex* wireless has become a hot research topic. Conventionally, most researchers held the view that it is generally not possible for radios (e.g., base stations, relays, or mobiles) to receive and transmit using the same frequency and time resources because the strong self-interference from the transmitter to the receiver will overwhelm the receiver circuitry, making it impossible to recover the incoming signal [1]. As a result, a long-held assumption in wireless system design is that radios can only operate in either half-duplex or out-of-band full-duplex mode, meaning that they transmit and receive either at different times, or over different frequency bands.

With recent advances in self-interference cancelation technologies, researches [2–8] have attempted to invalidate the above assumption by showing the feasibility of in-band full-duplex wireless, which allows radios to receive and transmit on the same frequency band simultaneously. The FD operation offers the potential to double the spectral efficiency in PHY layer, and thus is of great interest for next-generation wireless networks.

Apart from the potential to double spectral efficiency in PHY layer, FD techniques can also help to solve some important problems in upper layers like the medium access control (MAC) layer. From the perspective of MAC layer, enabling frame level FD transmissions, where a terminal is able to reliably receive an incoming frame while simultaneously transmitting an outgoing frame, can provide terminals with new capabilities. For example, terminals can detect collisions while transmitting in a contention-based network or receive instantaneous feedback from other terminals [9].

© The Author(s) 2016
Y. Liao et al., *Listen and Talk*, SpringerBriefs in Electrical and Computer Engineering, DOI 10.1007/978-3-319-33979-5_1

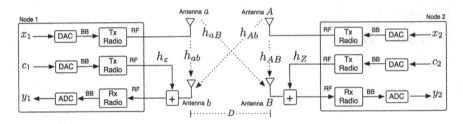

Fig. 1.1 RF model for a FD system

1.1.1 Self-Interference Cancelation

Despite the above attractive advantages of FD transmission, it has, until now, not seen widespread use due to the potential debilitating effects of self-interference (SI). Self-interference refers to the interference that a transmitting FD terminal causes to itself, which interferes with the desired signal being received by that terminal (Fig. 1.1).

To appreciate the impact of self-interference, consider the following example: femto base stations and mobile handsets transmit at 21 dBm, with a receiver noise floor of −100 dBm. If we assume 15 dB isolation between the base station's transmit and receive signal paths, then the base station's self-interference will be $21 - 15 - (-100) = 106$ dB above the noise floor. Thus, for a full-duplex base station to achieve the link SNR equal to that of a half-duplex counterpart, it must suppress self-interference by more than 106 dB, which is a daunting amount. On the other hand, the simple and intuitive way that FD terminals simply receive the mix of desired signal and self-interference first, and then subtract its own transmitted signal from the received signal does not work due to two main reasons: (1) the magnitude of the interfering signal can be large enough to saturate the receiver front end; and (2) even if there is no receiver front end saturation, the magnitude of the interfering signal at the input of the analog to digital converter (ADC) is much larger than the magnitude of the signal of interest. This results in quantization noise, which may be even larger than the signal of interest.

There are typically three classes of SI cancelation approaches, namely propagation-domain, analog-circuit-domain, and digital-domain approaches [9]. We will introduce them one by one in the rest of this section.

1.1.1.1 Propagation-Domain Self-Interference Suppression

Wireless-propagation-domain isolation techniques aim to electromagnetically isolate the transmit chain from the receive chain, i.e., to suppress the self-interference before it manifests in the receive chain circuitry, so that the downstream receiver hardware does not need to accurately process signals with a huge dynamic range.

In separate-antenna systems, the path loss between the FD terminals' transmit and receive antennas (or antenna arrays) can be increased by spacing them apart and/or by placing absorptive shielding between them, as quantified in [2, 6, 7]. Although the simplicity makes the pass-loss-based approaches attractive, their effectiveness is greatly limited by the size of device: the smaller the device, the less room there is to implement such techniques. Cross-polarization technique offers an additional mechanism to electromagnetically isolate the transmit and receive antennas. For example, an FD terminal can be sophistically designed so that it transmits only horizontally polarized signals and receives only vertically polarized signals, and the interference between them can be avoided [6, 10]. Similarly, with directional transmit and/or receive antennas (i.e., antennas with non-uniform radiation/sensing patterns), one may align their null directions to achieve the same goal [5]. In fact, by carefully placing a single receive antenna at precisely a location where the carrier waveforms are exactly 180° out of phase, one can near-perfectly null the received signal at the receive antenna, and the self-interference can be near-perfectly canceled [3, 10].

1.1.1.2 Analog-Circuit-Domain Self-Interference Cancellation

Analog-circuit-domain cancellation focuses on suppression of self-interference in the analog receive-chain circuitry before the ADC. This suppression may occur either before or after the downconverter and the LNA. In these techniques, a signal that resembles the SI at the receive-chain is generated, and utilized to cancel SI by signal subtraction. The generated signal can either be tapped at the transmit antenna, or obtained from the digital domain, and apply the necessary gain/phase/delay adjustments digitally (where it is much easier to do so), and then convert it to the analog-circuit domain for use in self-interference cancellation [11, 12].

For the suppression schemes before the downconverter, i.e., at RF, the canceling signal also needs to be upconverted to RF. We classify the analog cancelers based on whether the canceling signal is generated by processing the self-interference signal prior to or after upconversion. Those cancelers where the canceling signal is generated by processing prior to upconversion are called pre-mixer cancelers, whereas cancelers where the canceling signal is generated by processing afterwards are called post-mixer cancelers [13]. Figure 1.2a shows the structure of a pre-mixer canceler with processing function $f(\cdot)$, in which $x[n]$ is the original signal, and h_I is the self-interference channel from the transmit-chain to the receive-chain. Figure 1.2b is the schematic of a post-mixer canceler with processing function $g(\cdot)$. Functions $f(\cdot)$ and $g(\cdot)$ are ideal if they can completely eliminate self-interference from the received signal.

An analog canceler where the canceling signal is generated in baseband and the cancelation occurs in the analog baseband is called a baseband analog canceler. Figure 1.3 shows a representation of the baseband analog canceler. In baseband analog cancelers, transmitted signal $x[n]$ is processed by function $s(\cdot)$, and added directly to the received signal to perform the cancelation.

Fig. 1.2 Schematics of the
(a) pre-mixer and
(b) post-mixer analog
cancelers

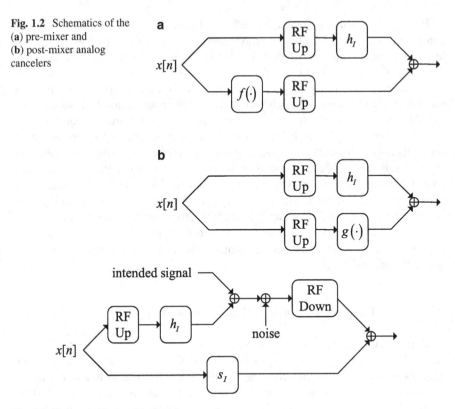

Fig. 1.3 Structure of a baseband analog canceler

1.1.1.3   Digital-Domain Self-Interference Cancellation

Digital-domain SI cancelation works in the digital domain after the received signal has been quantized by the ADC by applying sophisticated DSP techniques to the received signal [9]. The advantage of using digital-domain SI cancellation techniques is the reduction of circuit complexity and power consumption. However, the dynamic range of the ADC imposes strict limit on the maximum amount of SI suppression, which means that to implement digital-domain methods, a sufficient amount of the SI suppression must be done before the ADC using other techniques like the propagation-domain and/or analog-circuit-domain methods described above. In this sense, we can take the digital-domain cancellation as the last step of defense against self-interference, where the goal is to cancel the self-interference left over from the propagation-domain and analog-circuit-domain approaches.

1.1.2 State-of-Art Development and Applications of Full-Duplex Radio

FD technology, with its key feature of enabling simultaneous transmission and reception and its potential of doubling the spectral efficiency in PHY layer, has ignited great interest in both academia and industry. To fully explore the potential of FD, how to improve the performance of SI cancelation is still a great challenge to PHY layer design. In addition, some high-layer and system-level protocols need to be redesigned to accommodate capability of simultaneous transmission and reception.

For the hardware design and SI cancelation schemes, there have been significant progress in the design and realization of SI cancelers as well as in understanding the impact of hardware limitations and other imperfections on system performance. Typically, there have been increasing number of works regarding the design of single antenna FD [8, 14–16], and novel antenna designs [17] and cancelation mechanism proposals [6, 9] that report around 100 dB SI cancelation in practical systems. For the circuit non-idealities, there have been works about analysis and reduction of the effect of nonlinear distortion in the transmit and receive chains [18–20], phase noise [13, 21], and quantization noise [22, 23].

The redesign of upper layer designs [24–29] are also eye-catching. Some of them assume the existence of a central controller, and design centralized access mechanism [29], while most of them focus on the design of distributed MAC protocols with special concerns of mitigation of collision [30] and inter-node interference [26, 31], improvement of spectrum efficiency and power efficiency [32].

Also, the application of FD in some key scenarios like cellular networks [33], D2D networks [34], and relay networks [17, 35–38] have also been investigated by several groups, and many interesting results have come out recently.

1.2 Cognitive Radio Preliminaries

The existing and new wireless technologies, such as smart phones, wireless computers, WiFi home and business networks are rapidly consuming radio spectrum. Unlike the wired Internet, the wireless world has a limited amount of links to distribute. The usage of radio spectrum resources and the regulation of radio emissions are coordinated by national regulatory bodies like the Federal Communications Commission (FCC). These bodies assign spectrum to licensed holders, also known as primary users, on a long-term basis for large geographical regions. However, a large portion of the assigned spectrum remains under-utilized [39]. The inefficient usage of the limited spectrum necessitates the development of dynamic spectrum access (DSA), where users who have no spectrum licenses, also known as secondary users, are allowed to use the licensed spectrum without causing harmful interference at the primary users.

Cognitive radio is the key enabling technology that enables next generation communication networks, also known as DSA networks, to utilize the spectrum more efficiently in an opportunistic fashion without interfering with the primary users. It is defined as a radio that can change its transmitter parameters according to the interactions with the environment in which it operates. It differs from conventional radio devices in that a cognitive users have cognitive capability and reconfigurability [40]. Cognitive capability refers to the ability to sense and gather information from the surrounding environment, such as information about transmission frequency, bandwidth, power, modulation, etc. With this capability, secondary users can identify the best available spectrum. Reconfigurability refers to the ability to rapidly adapt the operational parameters according to the sensed information in order to achieve the optimal performance. By exploiting the spectrum in an opportunistic fashion, cognitive radio enables secondary users to sense which portion of the spectrum are available, select the best available channel, coordinate spectrum access with other users, and vacate the channel when a primary user reclaims the spectrum usage right [41].

1.2.1 Spectrum Sensing

Spectrum sensing enables the capability of a CR to measure, learn, and be aware of the radio's operating environment, such as the spectrum availability and interference status. When a certain frequency band is detected as not being used by the primary licensed user of the band at a particular time in a particular position, secondary users can utilize the spectrum, i.e., there exists a spectrum opportunity (Fig. 1.4).

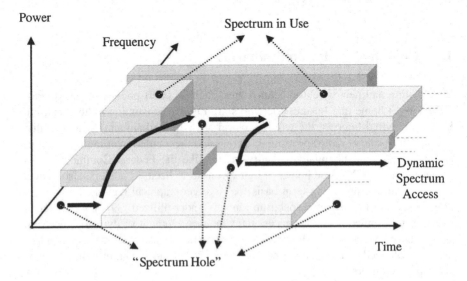

Fig. 1.4 The concept of spectrum holes

A wealth of literature on spectrum sensing focuses on primary transmitter detection based on the local measurements of secondary users, since detecting the primary users that are receiving data is in general very difficult. According to the priori information they require and the resulting complexity and accuracy, spectrum sensing techniques can be categorized in the following three types, namely energy detector, feature detector, and matched filter detector.

1.2.1.1 Energy Detector

Energy detection is the most common type of spectrum sensing because it is easy to implement and requires only basic information of the primary signal. The hypothesis model of the received signal is

$$
y(t) = \begin{cases} hx(t) + u(t), & \mathcal{H}_1, \\ u(t), & \mathcal{H}_0, \end{cases} \tag{1.1}
$$

where $x(t)$ is the primary user's signal, $u(t)$ is the additive white Gaussian noise (AWGN), and h is the channel gain from the primary transmitter to the secondary user. $\mathcal{H}_0$ is a null hypothesis, meaning there is no primary user present in the band, while $\mathcal{H}_1$ indicates the primary user's presence. The detection statistics of the energy detector O is the average energy of a given number of observed samples. The decision on whether the spectrum is occupied by the primary user is made by comparing the detection statistics O with a predetermined threshold ϵ.

Besides its low computational and implementation complexity and short detection time, there also exist some challenges in designing a good energy detector. For example, to achieve high accuracy in sensing, i.e., obtain low false alarm and miss detection probabilities, the sensing duration [42] and detection threshold [43] require careful design and optimization.

1.2.1.2 Feature Detector

In many cases, there are specific features, especially cyclostationary features associated with the primary user's signal, based on which a detector can distinguish cyclostationary signals from stationary noise [44]. As in most communication systems, the transmitted signals are modulated signals coupled with sine wave carriers, pulse trains, hopping sequences, or cyclic prefixes, while the additive noise is generally wide-sense stationary with no correlation, cyclostationary feature detectors can be utilized to differentiate noise from primary users' signal [45] and distinguish among different types of transmissions and primary systems [46].

Different from an energy detector which uses time-domain signal energy as test statistics, a cyclostationary feature detector performs a transformation from the time-domain into the frequency feature domain, in which the hypothesis test is conducted. We use the cyclic spectrum density (CSD) function for hypothesis test, which can be expressed as

$$S(f, \alpha) = \sum_{\tau=-\infty}^{\tau=\infty} R_y^\alpha (\tau) e^{-j2\pi f\tau}, \tag{1.2}$$

where $R_y^\alpha (\tau) = \mathbb{E}\left[y(t+\tau)y^*(t-\tau)e^{j2\pi ft}\right]$ is the cyclic autocorrelation function (CAF) of the received signal, and α is the cyclic frequency.

The CSD function has peaks when the cyclic frequency α equals to the fundamental frequencies of the transmitted signal. Under hypothesis $\mathcal{H}_0$, the CSD function does not have any peaks since the noise is non-cyclostationary signals. A peak detector [47] or a generalized likelihood ratio test [45, 46] can be further used to distinguish among the two hypothesis. Different primary communication systems using different air interfaces (modulation, multiplexing, coding, etc.) can also be differentiated by their different properties of cyclostationarity.

1.2.1.3   Matched Filter Detector

If secondary users know information about a primary user's signal a priori, then the optimal detection method is the matched filtering, since a matched filter can correlate the already known primary signal with the received signal to detect the presence of the primary user and thus maximize the SNR in the presence of additive stochastic noise. The merit of matched filtering is the short time it requires to achieve a certain detection performance such as a low probability of missed detection and false alarm, since a matched filter needs less received signal samples. However, its implementation complexity and power consumption is too high [48], because the matched filter needs receivers for all types of signals and corresponding receiver algorithms to be executed. Also, matched filtering requires perfect knowledge of the primary user's signal, such as the operating frequency, bandwidth, modulation type and order, pulse shape, packet format, etc, which is nearly impossible in many cases. If wrong information is used for matched filtering, the detection performance will be degraded a lot.

1.2.2   Dynamic Spectrum Allocation and Spectrum Sharing

Spectrum sharing among the secondary users and primary users in licensed spectrum bands is referred to as hierarchical access model [49] or licensed spectrum sharing. Primary users, usually not equipped with CR, have priority in using the spectrum band. Whenever they reclaim the spectrum usage, secondary users have to adjust their operating parameters, such as power, frequency, and bandwidth, to avoid interrupting the primary users.

With the detection techniques introduced in the previous section, secondary users are able to obtain an accurate estimation of the interference temperature or spectrum occupancy status. After this estimation, secondary users need to address issues such as when and how to use a spectrum band, how to co-exist with primary users and other secondary users, and which spectrum band they should sense and access if the current one in use is not available, which are in the scope of spectrum allocation and sharing strategies.

For secondary users, there are mainly two methods to perform spectrum access, namely spectrum underlay and spectrum overlay:

1. *Spectrum underlay:* In spectrum underlay, secondary users are allowed to transmit their data in the licensed spectrum band when primary users are also transmitting. The interference temperature model is imposed on secondary users' transmission power so that the interference at a primary user's receiver is within the interference temperature limit and primary users can deliver their packet to the receiver successfully. Spread spectrum techniques are usually adopted by secondary users to fully utilize the wide range of spectrum. However, due to the constraints on transmission power, secondary users can only achieve short-range communication. If primary users transmit data all the time in a constant mode, spectrum underlay does not require secondary users to perform spectrum detection to find available spectrum band.

2. *Spectrum overlay:* Spectrum overlay is also referred to as opportunistic spectrum access. Unlike spectrum underlay, secondary users in spectrum overlay will only use the licensed spectrum when primary users are not transmitting, so there is no interference temperature limit imposed on secondary users' transmission. Instead, secondary users need to sense the licensed frequency band and detect the spectrum white space, in order to avoid harmful interference to primary users.

In this book, we only focus on spectrum overlay technique, where secondary users need to precisely detect the location of a spectrum hole to perform secondary transmission, and the term "cognitive radio" refers to overlay cognitive radio in the rest of this book.

Regarding the network architecture, i.e., the existence of any central controllers, there are two types of spectrum sharing mechanism: centralized and distributed spectrum sharing [40]. When there exists a central entity that controls and coordinates the spectrum allocation and access of secondary users, then the spectrum allocation is centralized. If there is no such a central controller, that kind of spectrum sharing belongs to distributed spectrum sharing. In distributed spectrum sharing, each secondary user makes its own decision about its spectrum access strategy, mainly based on local observation of the spectrum dynamics.

From the perspective of the access behavior and interaction of secondary users, the spectrum allocation can also be divided into non-cooperative and cooperative manner [40]. Secondary users operate in cooperative manner when all secondary users work towards a common goal or coordinate their allocation and access in order to maximize their social welfare. On the other hand, in some cases, users are selfish in that they pursue their own benefit, they will choose their own access strategies without cooperate with each other. This is called non-cooperative manner.

1.2.3 Listen-Before-Talk Protocol

Most existing work on overlay cognitive radio networks (CRNs) employs the so-called "listen-before-talk" (LBT) protocol by half-duplex (HD) radio as illustrated in Fig. 1.5 [42, 50–53]. In the LBT protocol, SUs sense the target channel periodically, and begin transmission once a spectrum hole is detected. For this protocol, the design of sensing duration and sensing interval are of great importance, since there exist a tradeoff between accurate sensing, which requires long sensing time and short sensing interval, and high throughput [42, 51, 54]. Besides the optimization of sensing and transmission duration, some other solutions such as the adaptive sensing and transmission duration [55] have also been proposed to achieve higher spectrum efficiency. Though the conventional HD based LBT protocol is proved to be effective, it actually dissipates the precious resources by employing a time-division duplexing, and thus, unavoidably suffers from two major problems.

The first is that the SUs have to sacrifice the transmitting time for spectrum sensing, and even if the spectrum hole is long and continuous, the data transmission need to be split into small discontinuous fractions. This leads to the time waste and inefficiency of spectrum usage. The second problem is that during SUs' transmission, the SUs cannot detect the change of PU's states. This means that on one hand, when the primary user reclaims the spectrum, the secondary users cannot backoff or adjust their transmission promptly, which may end up with harmful interference to the primary transmission. On the other hand, when the primary user quits the spectrum, the secondary users may not be able to access immediately since they are not sensing the spectrum at that specific time, and there will be a fraction of spectrum hole remains unused.

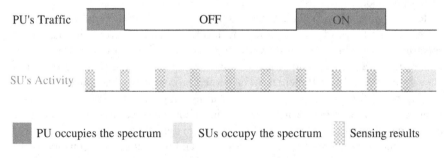

Fig. 1.5 Illustration of the listen-before-talk protocol

1.3   CSMA Basics

In a multi-user wireless communication system, the transmitting nodes share the limited spectrum resources. One critical issue is how to allocate theses resources to the nodes so that the nodes can access the medium fairly while collisions brought by concurrent transmissions can be prevented. Multiple access technique is the key solution to this issue. Multiple access methods can be divided into two main groups, namely contention-free channel access and contention-based random access methods. The contention-free channel access schemes coordinate shared access to avoid collision. There may be some central entities, e.g., the BSs, that control the access behavior of all the users; or there may be some common beacons or polling strategies that tell each node whether to access. On the contrary, in random access protocols, users compete to access the medium channel without any specific pre-assignment of the channel. There are mainly two kinds of random access protocols: one is carrier-sense-based mechanisms, represented by carrier sense multiple access (CSMA) mechanism; the other is non-carrier-sense-based mechanism, represented by ALOHA and Slotted ALOHA protocols.

In this section, we will focus on CSMA mechanism, which is one of most popular random access protocols in practice. We will briefly introduce the basic CSMA, CSMA with collision detection (CSMA/CD), and CSMA with collision avoidance (CSMA/CA) mechanisms.

1.3.1   Basic CSMA Mechanism

The key feature of CSMA is that each link with a pair of transmitter and receiver senses the medium before attempting transmission. A backoff algorithm based on Contention Window (CW) and Persistence Probability are used in CSMA to avoid simultaneous access to the channel. In the backoff algorithm, each node waits for a random time, limited to its CW before transmission. In the Persistence mechanism, each node maintains a persistence probability. Whenever it finds the channel free, it will access the channel with this probability [56]. Based on the persistency after the channel is sensed idle, there are typically the following three types of CSMA:

1. Non-Persistent CSMA: When the channel is found idle by any active node, the node sends its data; otherwise it waits for a random period and repeats the procedure again.
2. p-Persistent CSMA: this method is proper for time slotted channels. Once the availability of a free channel is detected by an active node, it sends its data with the probability of p or postpones its transmission until the next time slot with the probability of $q = 1 - p$. If the channel is busy, the station waits for the next time slot and repeats the process again.

3. 1-Persistent CSMA: In this method, if the channel is sensed free, data is sent instantaneously. Otherwise, the node waits until the channel is free. If there are multiple users with heavy load adopting the 1-Persistent CSMA scheme, there may be frequent collision.

1.3.1.1 CSMA/CD and CSMA/CA

The CSMA/CD and CSMA/CA are two similar enhanced versions of CSMA. In Ethernet, CSMA/CD is the basic of the MAC protocols, in which the transmitter simultaneously transmits and listens on the wired channel. Once a collision is detected, the transmitter aborts its transmission almost instantaneously. In this way, the channel utilization efficiency improves significantly compared to the basic CSMA since the remainder of the packet under collision does not need to be transmitted unnecessarily. Instead, the channel can be released for other productive transmissions. In addition, the collision can be mitigated or avoided by expanding the retransmission interval (i.e., backoff period) for the node to wait before a new transmission.

CSMA/CD has shown satisfying performance in wired networks. However, in wireless environment, CSMA/CD cannot be directly used because it is commonly assumed that wireless devices are unable to listen to the channel for collisions while transmitting. Thus, CSMA/CA has been proposed as an adaptation for wireless devices.

CSMA/CA, which is adopted as the MAC protocol in the IEEE 802.11 DCF, is very similar in operation to the CSMA/CD. In both protocols, the availability of the transmission medium is detected through carrier sensing. The distinguishing feature between CSMA/CA and CSMA/CD is that the former requires the receiver to send a positive acknowledgment (ACK) back to the transmitter if a frame is received correctly. Retransmission is scheduled by the transmitter if no ACK is returned. In CSMA/CD, the transmitter makes use of collision detection to determine whether a data frame has been transmitted successfully. The function of the MAC protocol is common to all three PHY layer options (i.e., DSSS, FHSS, DFIR) and is independent of the data rates [57].

In the CSMA/CA, every active node which has a new packet to transmit monitors channel activity before it accesses to the medium. If the channel is free for a common DCF Inter-Frame Spacing (DIFS) time, the node will start sending data. Otherwise, it will persist on monitoring until the channel is free for DIFS time. Next, backoff random time will be selected by the node based on the following equation:

$$\text{Backoff Time} = \text{Rand}\,(CW) \times \text{aSlotTime}, \qquad (1.3)$$

where CW is the contention window, and aSlotTime is the length of a time slot. Collision is still possible in this protocol due to concurrent transmission. After each unsuccessful transmission, CW is multiplied by σ (which has the default value of 2), which is called the persistence coefficient and the backoff time selection

Fig. 1.6 An example of exponential increase of *CW*

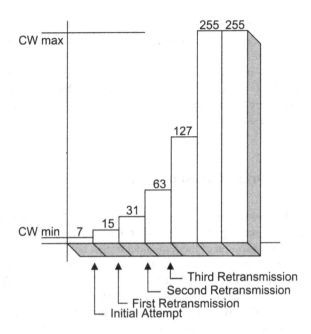

and transmission process will be repeated again. The process goes on till the packet is transmitted successfully or the size of the contention window reaches its maximum value $CW_{max} = \sigma^m CW_{min}$, i.e., after each failed transmission, $CW \leftarrow \min\{\sigma CW, CW_{max}\}$. After a packet is successfully delivered, the CW will be set back to CW_{min} (Fig. 1.6).

1.3.2 Wireless CSMA/CD

MAC protocols in wired LANs are based on the principles of CSMA/CD, in which the transmitter simultaneously listens and transmits on the wired channel. On detecting a collision, the transmitter aborts its own transmission almost instantaneously. Wireless MAC protocols, nowadays, however, must rely on CSMA/CA, in which transmitters must finish the entire packet and then infer a collision from the absence of an ACK from their receivers. Channel utilization degrades due to the retransmission of the failed packets. To reduce channel waste, it would be desirable to emulate CSMA/CD-like behavior in wireless networks.

Over the past years, some trials have come out to mitigate this problem by approximating CSMA/CD in half-duplex systems [58, 59]. The protocol designed in [58] requires the users to randomly perform additional sensing while they are transmitting, which leads to additional discontinuity of the transmission. In [59], the collision is detected by the receiver, and a notification signal is sent back to the transmitter with unique signature. However, the performance of collision detection at the receiver side as well as the performance of notification detection at the transmitter still remains questionable.

Thus, in Chap. 4, we manage to introduce FD techniques into multiple access networks. With FD capabilities, users are able to sense the channel concurrently when they are transmitting, which provides the possibility of realizing CSMA/CD in wireless systems.

1.4   Organization of the Book

In this book, we mainly present the idea of "listen-and-talk", i.e., enabling simultaneous sensing and transmission in different networks by full-duplex techniques. The organization of the remainder of this book is as follows.

In Chap. 2, we explore the FD techniques in a simple CRN with one PU and one pair of SUs, and present a novel "Listen-and-Talk" (LAT) protocol, by which the SUs can simultaneously perform spectrum sensing and data transmission. Specifically, by equipping with FD radios, SUs can sense the target spectrum band continuously, and determine if the PUs are busy or idle in every short slot, which guarantees that SUs can react promptly to the PU's state change.

Based on the basic LAT protocol that evolves only one pair of SUs, in Chap. 3, we consider more complicated networks. First, some important issues in the cooperative spectrum sensing under the LAT protocol are addressed in Sect. 3.1. In the cooperation, the major difference from the conventional half-duplex scenarios is that the interference from the transmitting SU to other cooperative SUs will significantly degrade others' local sensing performance. Taken this feature into consideration, a feasible cooperation scheme is provided and proved effective in Sect. 3.1. Then, we consider FD cognitive networks with multiple contending users. Both distributed and centralized spectrum sharing and resource allocation mechanisms are discussed, and analyzed in Sect. 3.2.

Apart from the FD cognitive networks, which is a representative in vertical spectrum sharing, in Chap. 4, we also manage to apply the idea of LAT into horizontal spectrum sharing, featured by the WiFi networks. For the FD WiFi networks, we provide a cross-layer protocol design based on the basic idea of simultaneous sensing and transmission.

Conclusions of the whole book is provided in Chap. 5. Also, we are aware that there still exist numerous challenging problems such as FD MIMO networks, and the coexistence problem of FD networks and other contentional networks, all of which wait to be investigated in the future work. We list some of the challenges at the end of Chap. 5, which can indeed provide feasible directions of employing FD techniques in future wireless communication networks.

References

1. A. Goldsmith, *Wireless Communications.* Cambridge, U.K.: Cambridge University Press, 2005.
2. M. Duarte and A. Sabharwal, "Full-duplex Wireless Communications Using Off-the-Shelf Radios: Feasibility and First Results," in *Proc. Asilomar Conf. on Signals, Systems and Computers*, pp. 1558–1562, Pacific Grove, CA, Nov. 2010.
3. J. Choi, J. Mayank, S. Kannan, L. Philip, and K. Sachin, "Achieving Single Channel, Full Duplex Wireless Communication," in *Proc. ACM MobiCom 2010*, Chicago, IL, Sep. 2010.
4. M. Jain, J. I. Choi, T. Kim, D. Bharadia, S. Seth, K. Srinivasan, P. Levis, S. Katti, and P. Sinha. "Practical, Real-time, Full Duplex Wireless," in *Proc. ACM MobiCom 2011*, New York, NY, Sep. 2011.
5. E. Everett, M. Duarte, C. Dick, and A. Sabharwal "Empowering Full-duplex Wireless Communication by Exploiting Directional Diversity," in *Proc. Asilomar Conf. on Signals, Systems and Computers*, pp. 2002–2006, Pacific Grove, CA, Nov. 2011.
6. E. Everett, A. Sahai, and A. Sabharwal, "Passive Self-Interference Suppression for Full-Duplex Infrastructure Nodes," *IEEE Trans. on Wireless Comm.*, vol. 13, no. 2, pp. 680–694, Feb. 2014.
7. A. Sahai, G. Patel, and A. Sabharwal, Pushing the Limits of Full-Duplex: Design and Real-Time Implementation, Rice University, Houston, TX, USA, Tech. Rep. TREE1104.
8. D. Bharadia, E. McMilin, and S. Katti, "Full Duplex Radios," in *Proc. ACM SIGCOMM.'13*, pp. 375–386, Hong Kong, China, Aug. 2013.
9. A. Sabharwal, P. Schniter, D. Guo, D. W. Bliss, S. Rangarajan, and R. Wichman, "In-Band Full-Duplex Wireless: Challenges and Opportunities," *IEEE Journal on Selected Areas in Communications*, vol. 32, no. 9, pp. 1637–1652, Sep. 2014.
10. E. Aryafar, M. Khojastepour, K. Sundaresan, S. Rangarajan, and M. Chiang, "MIDU: Enabling MIMO Full Duplex," in *Proc. ACM MobiCom'12*, pp. 257–268, Istanbul, Turkey, Aug. 2012.
11. M. Duarte, C. Dick, and A. Sabharwal, "Experiment-driven Characterization of Full-duplex Wireless Systems," *IEEE Trans. Wireless Commun.*, vol. 11, no. 12, pp. 4296–4307, Dec. 2012.
12. M. Duarte, A. Sabharwal, V. Aggarwal, R. Jana, K. K. Ramakrishnan, C. W. Rice, and N. K. Shankaranarayanan, "Design and Characterization of a Full-duplex Multiantenna System for WiFi Networks," *IEEE Trans. Veh. Commun.*, vol. 63, no. 3, pp. 1160–1177, Mar. 2014.
13. A. Sahai, G. Patel, C. Dick, and A. Sabharwal, "On the Impact of Phase Noise on Active Cancelation in Wireless Full-Duplex," *IEEE Trans. on Vehi. Tech.*, vol. 62, no. 9, pp. 4494–4510, Nov. 2013.
14. L. Laughlin, M. A. Beach, K. A. Morris, and J. L. Haine, "Optimum Single Antenna Full Duplex Using Hybrid Junctions," *IEEE Journal on Selected Areas in Comm.*, vol. 32, no. 9, pp. 1653–1661, Sep. 2014.
15. N. Phungamngern, P. Uthansakul, and M. Uthansakul, "Digital and RF Interference Cancellation for Single-Channel Full-duplex Transceiver Using a Single Antenna," in *Proc. IEEE 10th Int. Conf. ECTI-CON*, pp. 1–5, Krabi, Thailand, May 2013.
16. M. E. Knox, "Single Antenna Full Duplex Communications using a Common Carrier," in *Proc. IEEE 13th Annu. Wireless and Microwave Tech. Conf.*, pp. 1–6, Cocoa Beach, FL, Apr. 2012.
17. M. Heino, D. Korpi, T. Huusari, E. Antonio-Rodriguez, S. Venkatasubramanian, T. Riihonen, L. Anttila, C. Icheln, K. Haneda, R. Wichman, and M. Valkama, "Recent Advances in Antenna Design and Interference Cancellation Algorithms for In-Band Full Duplex Relays," *IEEE Comm. Magazine*, vol. 53, no. 5, pp. 91–101, May 2015.
18. D. Korpi, T. Riihonen, V. Syrjälä, L. Anttila, M. Valkama, and Ri. Wichman, "Full-Duplex Transceiver System Calculations: Analysis of ADC and Linearity Challenges," *IEEE Trans. on Wireless Comm.*, vol. 13, no. 7, pp. 3821–3836, Jul. 2014.
19. E. Ahmed, A. Eltawil, and A. Sabharwal, "Self-interference Cancellation with Nonlinear Distortion Suppression for Full-duplex Systems," in *Proc. 47th Asilomar Conf. on Signals, Syst. Comput.*, pp. 1199–1203, Pacific Grove, CA, Nov. 2013.

20. D. Korpi, L. Anttila, V. Syrjälä, and M. Valkama, "Widely Linear Digital Self-Interference Cancellation in Direct-Conversion Full-Duplex Transceiver," *IEEE Journal on Selected Areas in Comm.*, vol 32, no. 9, pp. 1674–1687, Sep. 2014.
21. V. Syrjälä, M. Valkama, L. Anttila, T. Riihonen, and D. Korpi, "Analysis of Oscillator Phase-Noise Effects on Self-interference Cancellation in Full-Duplex OFDM Radio Transceivers," *IEEE Trans. on Wireless Comm.*, vol. 13, no. 6, pp. 2977–2990, Jun. 2014.
22. T. Riihonen and R. Wichman, "Analog and Digital Self-interference Cancellation in Full-duplex MIMO-OFDM Transceivers with Limited Resolution in A/D Conversion," in *Proc. 46th Asilomar Conf. on Signals, Syst. Comput.*, pp. 45–49, Nov. 2012.
23. E. Ahmed, A. Eltawil, and A. Sabharwal, "Rate Gain Region and Design Tradeoffs for Full-duplex Wireless Communications," *IEEE Trans. on Wireless Comm.*, vol. 12, no. 7, pp. 3556–3565, Jul. 2013.
24. Y. Liao, K. Bian, L. Song, and Z. Han, "Full-duplex MAC Protocol Design and Analysis," *IEEE Comm. Letters*, vol. 19, no. 7, pp. 1185–1188, Jul. 2015.
25. K. M. Thilina, H. Tabassum, E. Hossain, and D. I. Kim, "Medium Access Control Design for Full Duplex Wireless Systems: Challenges and Approaches," *IEEE Comm. Magazine*, vol. 53, no. 5, pp. 112–120, May 2015.
26. S. Goyal, P. Liu, O. Gurbuz, E. Erkip, and S. Panwar, "A Distributed MAC Protocol for Full Duplex Radio," in *Proc. 47th Asilomar Conf. on Signals, Syst. Comput.*, pp. 788–792, Nov. 2013.
27. Y. Zhang, L. Lazos, K. Chen, B. Hu, and S. Shivaramaiah, "FD-MMAC: Combating Multi-Channel Hidden and Exposed Terminals Using a Single Transceiver," in *Proc. IEEE Infocom*, pp. 2742–2750, Toronto, ON, Apr. 2014.
28. K. Tamaki, A. Raptino H., Y. Sugiyama, M. Bandai, S. Saruwatari, and T. Watanabe, "Full Duplex Media Access Control for Wireless Multi-hop Networks," in *Proc. IEEE 7th VTC-Spring*, pp. 1–5, Dresden Germany, Jun. 2013.
29. J. Y. Kim, O. Mashayekhi, H. Qu, M. Kazadiieva, and P. Levis, "JANUS: A Novel MAC Protocol for Full Duplex Radio," CSTR, 2013.
30. W. Cheng, X. Zhang, and H. Zhang, "RTS/FCTS Mechanism Based Full-Duplex MAC Protocol for Wireless Networks," in *Proc. IEEE Globecom*, pp. 5017–5022, Atlanta, GA, Dec. 2013.
31. Y. Sugiyama, K. Tamaki, S. Saruwatari, and T. Watanabe, "A Wireless Full-duplex and Multi-hop Network with Collision Avoidance using Directional Antennas," in *Proc. 7th Int'l Conf. on Mobile Computing and Ubiquitous Networking (ICMU)*, pp. 38–43, Singapore, Jan. 2014.
32. S. Kim and W. E. Stark, "On the Performance of Full Duplex Wireless Networks," in *Proc. 47th Annual Conf. on Information Sciences and Systems (CISS)*, pp. 1–6, Baltimore, MD, Mar. 2013.
33. S. Goyal, P. Liu, S. S Panwar, R. A. DiFazio, R. Yang, and E. Bala, "Full Duplex Cellular Systems: Will Doubling Interference Prevent Doubling Capacity?" *IEEE Comm. Magazine*, vol. 53, no. 5, pp. 121–127, May 2015.
34. L. Wang, F. Tian, T. Svensson, D. Feng, M. Song, and S. Li, "Exploiting Full Duplex for Device-to-Device Communications in Heterogeneous Networks," *IEEE Comm. Magazine*, vol. 53, no. 5, pp. 146–152, May 2015.
35. H. Cui, L. Song, and B. Jiao, "Multi-Pair Two-Way Amplify-and-Forward Relaying with Very Large Number of Relay Antennas," *IEEE Trans. on Wireless Comm.*, vol. 13, no. 5, pp. 2636–2645, May 2014.
36. H. Cui, M. Ma, L. Song, and B. Jiao, "Relay Selection for Two-Way Full Duplex Relay Networks with Amplify-and-Forward Protocol," *IEEE Trans. on Wireless Comm.*, vol. 13, no. 7, pp. 3768–3777, Jul. 2014.
37. K. Yang, H. Cui, L. Song, and Y. Li, "Efficient Full-Duplex Relaying with Joint Antenna-Relay Selection and Self-Interference Suppression," *IEEE Trans. on Wireless Comm.*, vol. 14, no. 7, pp. 3991–4005, Jul. 2015.
38. Z. Zhang, X. Chai, K. Long, A. V. Vasilakos, and L. Hanzo, "Full Duplex Techniques for 5G Networks: Self-Interference Cancellation, Protocol Design, and Relay Selection," *IEEE Comm. Magazine*, vol. 53, no. 5, pp. 128–137, May 2015.

39. Federal Communications Commission, "Spectrum Policy Task Force", Rep. ET Docket no. 02–135, Nov. 2002.
40. I. F. Akyildiz, W. Y. Lee, M. C. Vuran, and S.Mohanty, "Next Generation/Dynamic Spectrum Access/Cognitive Radio Wireless Networks: A Survey," *Computer Networks*, vol. 50, no. 13, pp. 2127–2159, Sep. 2006.
41. B. Wang and K. J. R. Liu, "Advances in Cognitive Radio Networks: A Survey," *IEEE Journal of Selected Topics in Signal Processing*, vol. 5, no. 1, pp. 5–23, Feb. 2011.
42. Y. C. Liang, Y. Zeng, E. C. Y. Peh, and A. T. Hoang, "Sensing-Throughput Tradeoff for Cognitive Nadio Networks," *IEEE Trans. on Wireless Comm.*, vol. 7, no. 4, pp. 1326–1337, Apr. 2008.
43. F. F. Digham, M.-S. Alouini, and M. K. Simon, "On the Energy Detection of Unknown Signals Over Fading Channels," *IEEE Trans. on Commum.*, vol. 55, no. 1, pp. 21–24, Jan. 2007.
44. W. A. Gardner, "Signal Interception: A Unifying Theoretical Framework for Feature Detection," *IEEE Trans. on Commun.*, vol. 36, no. 8, pp. 897–906, Aug. 1988.
45. M. Öner and F. Jondral, "Air Interface Recognition for a Software Radio System Exploiting Cyclostationarity," in *Proc. 15th IEEE Int. Symp. on Personal, Indoor, Mobile Radio Commun. (PIMRC)*, vol. 3, pp. 1947–1951, Sep. 2004.
46. J. Lunden, V. Koivunen, A. Huttunen, and H. Poor, "Spectrum Sensing in Cognitive Radios Based on Multiple Cyclic Frequencies," in *Proc. 2nd Int. Conf. Cognitive Radio Oriented Wireless Netw. Commun. (CrownCom)*, pp. 37–43, Orlando, FL, Aug. 2007.
47. L. P. Goh, Z. Lei, and F. Chin, "Feature Detector for DVB-T Signal in Multipath Fading Channel," in *Proc. 2nd Int. Conf. Cognitive Radio Oriented Wireless Netw. Commun. (CrownCom)*, pp. 234–240, Orlando, FL, Aug. 2007.
48. D. Cabric, S. Mishra, and R. Brodersen, "Implementation Issues in Spectrum Sensing for Cognitive Radios," in *Proc. 38th Asilomar Conf. on Signals, Syst. and Comput.*, vol. 1, pp. 772–776, Nov. 2004.
49. Q. Zhao and B. Sadler, "A Survey of Dynamic Spectrum Access," *IEEE Signal Process. Mag.*, vol. 24, no. 3, pp. 79–89, May 2007.
50. T. Yucek and H. Arslan, "A Survey of Spectrum Sensing Algorithms for Cognitive Radio Applications," *IEEE Communications Surveys & Tutorials*, vol. 11, no. 1, pp. 116–130, Mar. 2009.
51. S. Huang, X. Liu, and Z. Ding, "Short Paper: On Optimal Sensing and Transmission Strategies for Dynamic Spectrum Access," in *Proc. IEEE DySPAN*, Chicago, IL, Oct. 2008.
52. S. Huang, X. Liu, and Z. Ding, "Opportunistic Spectrum Access in Cognitive Radio Networks," in *Proc. IEEE INFOCOM 2009*, Rio de Janeiro, Brazil, Apr. 2009.
53. Q. Zhao, L. Tong, A. Swami, and Y. Chen, "Decentralized Cognitive MAC for Opportunistic Spectrum Access in Ad Hoc Networks: A POMDP Framework," *IEEE Journal on Selected Areas in Comm.*, vol. 25, no. 3, pp. 589–600, Apr. 2007.
54. Q. Zhao, S. Geirhofer, L. Tong, and B. M. Sadler, "Optimal Dynamic Spectrum Access Via Periodic Channel Sensing," in *IEEE Wireless Comm. and Networking Conf (WCNC) 2007*, pp. 33–37, Hongkong, China, Mar. 2007.
55. W. Afifi, A. Sultan and M. Nafie, "Adaptive Sensing and Transmission Durations for Cognitive Radios," in *Proc. IEEE Int'l Symp. on Dynamic Spectrum Access Networks (DySPAN)*, pp. 380–388, Aachen, Germany, May 2011.
56. M. Ghazvini, N. Movahedinia, K. Jamshidi, and N. Moghim, "Game Theory Applications in CSMA Methods," *IEEE Comms. Surveys & Tutorials*, Vol. 15, No. 3, Third Quarter 2013.
57. B. Bing, "Measured Performance of the IEEE 802.11 Wireless LAN," in *IEEE Conf. on Local Computer Networks*, pp. 34–42, Lowell, MA, Oct. 1999.
58. H.-H. Choi, J.-M. Moon, I.-H. Lee, and H. Lee, "Carrier Sense Multiple Access with Collision Resolution," *IEEE Comms. Letters*, vol. 17, no. 6, pp. 1284–1287, Jun. 2013.
59. S. Sen, R. R. Choudhury, and S. Nelakuditi, "CSMA/CN: Carrier Sense Multiple Access With Collision Notification," *IEEE Trans. on Netw.*, vol. 20, no. 2, pp. 544–556, Apr. 2012.

Chapter 2
Full-Duplex Cognitive Radio Networks

With the rapid growth of demand for ever-increasing data rate, spectrum resources have become increasingly scarce. However, an early study by FCC shows that most of the allocated spectrum is largely under-utilized in vast temporal and geographic dimensions [1]. Cognitive radio (CR), as a promising solution to spectrum reuse, has caused wide attention for more than a decade [2, 3]. In cognitive radio networks (CRNs), unlicensed or secondary users (SUs) are allowed to opportunistically utilize the vacant slots in the spectrum allocated to primary users (PUs). SUs therefore need to search for spectrum holes reliably and efficiently to protect the PU networks as well as maximize their own throughput [4].

Traditionally, the so-called "listen-before-talk" (LBT) strategy in which SUs sense the target channel before transmission has been extensively studied [5]. In this strategy, the design of sensing and transmission period is crucial to the improvement of secondary throughput [6] and [7]. The LBT strategy requires little infrastructure support and it proves to be effective. However, it still has two major problems:

1. The SUs have to sacrifice transmission time for spectrum sensing, and even if the spectrum hole is long and continuous, the data transmission needs to be split into small discontinuous slots;
2. During SUs' transmission, SUs cannot detect the changes of PUs' states, which leads to collision when PUs arrive and the spectrum waste when PUs leave.

The intrinsic reason is that most current deployed radios for wireless communications are half-duplex such that to dissipate the precious resources by either employing time-division or frequency-division.

Nowadays, the rapid development of full-duplex (FD) communications has shown the possibility of simultaneous transmission and reception on the same frequency band in wireless devices. Motivated by the realization of FD transmission technique, we try to explore a new way to bypass the above problems by introducing FD SUs into CRNs.

© The Author(s) 2016
Y. Liao et al., *Listen and Talk*, SpringerBriefs in Electrical and Computer Engineering, DOI 10.1007/978-3-319-33979-5_2

In this chapter, we present a "listen-and-talk" (LAT) protocol that allows SUs to simultaneously perform spectrum sensing and data transmission [8, 9]. We assume that the PU can change its state at any time and each SU is equipped with two antennas for simplicity and without the loss of generality. Specifically, at each moment, one of the antennas at each SU senses the target spectrum band, and judges if the PU is busy or idle; while the other antenna transmits data simultaneously or keeps silent on basis of the sensing results.

Apparently, the proposed FD-CR system is totally different from the traditional HD based one in many aspects, including:

- Spectrum sensing: in FD CRNs, sensing is continuous, but the received signal for sensing is interfered by the residual self-interference (RSI), which degrades the signal-to-interference-plus-noise ratio (SINR) in sensing. While in HD CRNs, there exists no RSI in received signal for sensing, but the sensing process is discontinuous and only takes a small fraction of each slot. This leads to unreliable sensing performance due to the inefficient number of samples to make decisions;
- Data transmission: in traditional HD CRNs, the SUs can only utilize the remaining part of each slot after sensing for data transmission. On contrary, in FD CRNs, SUs can continuously transmit as long as PUs are absent. However, in FD CRNs, the data transmission affects the sensing process, and thus there exists a constraint of transmit power to achieve acceptable sensing performance.

As shown above, the FD technology enables to explore another dimension of the network resources for increasing the capacity of CRNs. This thus requires a new design of signal processing techniques, resource allocation algorithms, and network protocols. For example, one of the major challenges faced by FD-CR is how to optimize the transmit power for the FD source node to maximize the system throughput.

This chapter comprehensively discusses the novel protocol design issues, key system parameter derivation, and practical algorithms for FD CRNs [8]. Specifically, the structure of this chapter is as follows. In Sect. 2.1, we present the basic system model with one PU and one pair of SUs. Then, in Sect. 2.2, the proposed LAT protocol is elaborated, and the key parameter design and performance analysis are addressed. At last, Sect. 2.3 provides a brief conclusion on this chapter.

2.1   System Model

In this section, the system model of the overall network is presented, and the concept of simultaneous sensing and transmission under imperfect self-interference suppression is elaborated.

Fig. 2.1 System model of
the LAT protocol

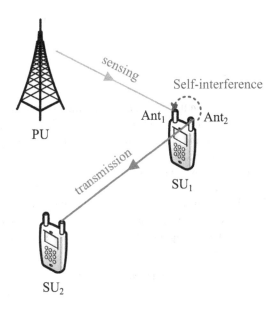

2.1.1 System Model

Consider a CR system consisting of one PU and one SU pair as shown in Fig. 2.1, in which SU_1 is the secondary transmitter and SU_2 is the receiver. Each SU is equipped with two antennas Ant_1 and Ant_2. Ant_1 is used for spectrum sensing, and Ant_2 is used for secondary communication. The transmitter SU_1 uses both Ant_1 and Ant_2 for simultaneous spectrum sensing and secondary transmission with the help of FD techniques, while the receiver SU_2 uses only Ant_2 to receive signal from SU_1.[1]

The spectrum band occupancy by the PU is modeled as an alternating busy/idle random process where the PU can access the spectrum at any time. We assume that the probabilities of the PU's arrival and departure remain the same across the time, and the holding time of either state is distributed as the exponential distribution [11]. We denote the variables of the idle period and busy period of the PU as t_0 and t_1, respectively. And let $\tau_0 = \mathbb{E}[t_0]$ and $\tau_1 = \mathbb{E}[t_1]$ represent the average idle and busy duration. According to the property of exponential distribution, the probability density functions (PDFs) of t_0 and t_1 can be written as, respectively,

---

[1] In this model, only the secondary transmitter (SU_1) performs spectrum sensing, while the receiver (SU_2) does not. This transmitter-only sensing mechanism is widely adopted in today's cognitive radio, considered that secondary transmitters and receivers cannot continuously exchange sensing results without interfering the primary network. Besides, we assume that SU_2 has two antennas for fairness and generality, since SU_2 does not always need to be the receiver.

$$f_0(t_0) = \frac{1}{\tau_0} e^{-\frac{t_0}{\tau_0}},$$

$$f_1(t_1) = \frac{1}{\tau_1} e^{-\frac{t_1}{\tau_1}}. \tag{2.1}$$

For SUs, on the other hand, only the idle period of the spectrum band is allowed to be utilized. To detect the spectrum holes and avoid collision with the PU, SU_1 needs to sample the spectrum at sampling frequency f_s, and make decisions of whether the PU is present after every N_s samples, This makes the secondary traffic time-slotted, with slot length $T = N_s/f_s$.

Considering the common case that f_s can be sufficiently high and the state of PU changes sufficiently slowly, we assume that $\tau_0, \tau_1 \gg T$ and N_s is a sufficiently large integer. If we divide the PU traffic into slots in accordance with SU's sensing process, the probability that PU changes its state in a stochastic slot can be derived as follows.

- The PU arrives in a stochastic slot:

$$\mu = \int_0^T f_0(t_0) \, dt_0 = 1 - e^{-\frac{1}{l_0}}, \tag{2.2}$$

where $l_0 = \tau_0/T$ and we assume it to be a large integer.
- The PU leaves in a stochastic slot:

$$\nu = \int_0^T f_1(t_1) \, dt_1 = 1 - e^{-\frac{1}{l_1}}, \tag{2.3}$$

where $l_1 = \tau_1/T$ is assumed to be a large integer.

Note that when l_0 and l_1 are sufficiently large, we have $\mu \approx \frac{1}{l_0}$ and $\nu \approx \frac{1}{l_1}$.

2.1.2 Simultaneous Sensing and Transmission

With the help of FD technique, SU_1 can detect the PU's presence when it is transmitting signal to SU_2. However, as shown by the dotted arrow in the system model in Fig. 2.1, the challenge of using FD technique is that the transmit signal at Ant_2 is received by Ant_1, which causes self-interference at Ant_1. Note that for Ant_1, the received signal is affected by the state of the transmit antenna (Ant_2): when Ant_2 is silent, the received signal at Ant_1 is free of self-interference, and the spectrum sensing is the same as the conventional half-duplex sensing. Thus, we consider the circumstances when SU_1 is transmitting or silent separately.

When SU_1 is silent, the received signal at Ant_1 is the combination of potential PU's signal and noise. The cases when the PU is busy or idle are referred to as hypothesis $\mathcal{H}_{01}$ and $\mathcal{H}_{00}$, respectively. The received signal at Ant_1 under each hypothesis can be written as

$$y = \begin{cases} h_s s_p + u, & \mathcal{H}_{01}, \\ u, & \mathcal{H}_{00}, \end{cases} \tag{2.4}$$

where s_p denotes the signal of the PU, h_s is the channel from the PU to Ant_1 of SU_1, and $u \sim \mathcal{CN}\left(0, \sigma_u^2\right)$ denotes the complex-valued Gaussian noise. Without loss of generality, in this book, we assume that s_p is PSK modulated with variance σ_p^2, and h_s is a Rayleigh channel with zero mean and variance σ_h^2. Note that other modulation modes of the PU's signal and channel conditions, such as Gaussian channels, Rician channels, or pathloss models for real conditions may lead to different distributions of the received signal for sensing, but they do not affect the main conclusion in the remainder of this book.

When SU_1 is transmitting to SU_2, RSI is introduced to the received signal at Ant_1. Thus, the received signal varies from (2.4), and can be written as

$$y = \begin{cases} h_s s_p + w + u, & \mathcal{H}_{11}, \\ w + u, & \mathcal{H}_{10}, \end{cases} \tag{2.5}$$

where $\mathcal{H}_{11}$ and $\mathcal{H}_{10}$ are the hypothesises under which the SU is transmitting and the PU is busy or idle, respectively. Variable w in (2.5) denotes the RSI at Ant_1, which can be modeled as the Rayleigh distribution with zero mean and variance $\chi^2 \sigma_s^2$ [6, 10], where σ_s^2 denotes the secondary transmit power at Ant_2 and χ^2 represents the degree of self-interference suppression, which is defined as

$$\chi^2 := \frac{\text{Power of the RSI}}{\text{Transmit power}}.$$

The parameter χ^2 is commonly expressed in dBs, and indicates how well can the self-interference be suppressed.

Spectrum sensing refers to the hypothesis test in either (2.4) or (2.5). Given that SU_1 has the information of its own state (silent or transmitting), it can automatically choose one pair of the hypothesises to test, i.e., when SU_1 is silent, it judges the PU's state by testing $\{\mathcal{H}_{00}, \mathcal{H}_{01}\}$; and when SU_1 is transmitting, it chooses $\{\mathcal{H}_{10}, \mathcal{H}_{11}\}$ to test. If $\mathcal{H}_{00}$ or $\mathcal{H}_{10}$ is verified, the spectrum is judged idle; otherwise, the spectrum is judged occupied by the PU.

2.2   Listen-and-Talk Protocol

In this section, we first present the proposed "Listen-and-Talk" (LAT) protocol, its key parameter design in spectrum sensing to meet the constraint of collision ratio to the primary network, and the analysis of its sensing performance and the secondary throughput. Then, the simple model shown in Fig. 2.1 with only one pair of SUs is extended to the scenario with multiple SU pairs, and the cooperation and contention among the SU pairs are considered.

2.2.1   Protocol Description

Figure 2.2 shows the sensing and spectrum access procedure of the LAT protocol. SU_1 performs sensing and transmission simultaneously by using the FD technique: Ant_1 senses the spectrum continuously while Ant_2 transmits data when a spectrum hole is detected. Specifically, SU_1 keeps sensing the spectrum with Ant_1 with sampling frequency f_s, which is shown in the line with down arrows. At the end of each slot with duration T, SU_1 combines all samples in the slot and makes the decision of the PU's presence. The decisions are represented by the small circles, in which the higher ones denote that the PU is judged active, while the lower ones

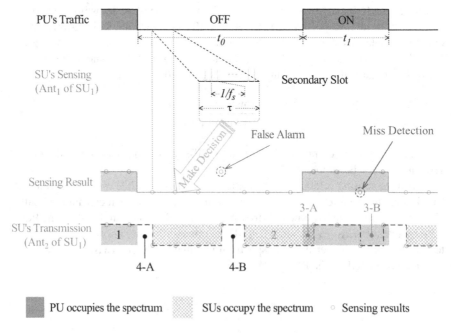

Fig. 2.2 The LAT protocol

denote otherwise. The activity of SU_1 is instructed by the sensing decisions, i.e., SU_1 can access the spectrum in the following slot when the PU is judged absent, and it needs to backoff otherwise.

With the LAT protocol, ideally, the spectrum hole can be fully utilized by SUs without any interference to the primary network due to the following characteristics of the LAT protocol:

- Secondary transmitters no longer need to sacrifice a fraction of each time slot to sense the spectrum silently. Instead, they can sense whether the PU arrives continuously while transmitting.
- Secondary transmitters can response to the PU's arrival and departure promptly since there is no "blind duration" as in the conventional half-duplex protocols, in which SUs either transmit or keep silent without sensing the occupancy of the spectrum.

However, energy detection requires minimum time duration, i.e., minimum slot length to make reliable decisions. Thus, SUs may not be able to detect the PU's state change immediately, which leads to potential collision and spectrum waste when the PU changes its state. Also, due to the sensing noise and the RSI, sensing errors may occur. There are two types of sensing errors, namely false alarm and miss detection. The former refers to the case when SUs judges that the spectrum is occupied by a PU when it is actually not, while the later means that SUs fail to detect the presence of the PU's signal.

From the discussion above, it can be seen that there exist the following four states of spectrum utilization:

- $State_1$: the spectrum is occupied only by the PU, and SU_1 is silent.
- $State_2$: the PU is absent, and SU_1 utilizes the spectrum.
- $State_3$: the PU and SU_1 both transmit, and a collision happens.
- $State_4$: neither the PU nor SU_1 is active, and there remains a spectrum hole.

Among these four states, $State_1$ and $State_2$ are the normal cases, and $State_3$ and $State_4$ are referred to as collision and spectrum waste, respectively. There are two reasons leading to $State_3$ and $State_4$: (A) the PU changes its state during a slot, and (B) sensing error, i.e., false alarm and miss detection.

2.2.1.1 Energy Detection

We adopt energy detection as the sensing scheme, in which the average received power in a slot is used as the test statistics O:

$$O = \frac{1}{N_s} \sum_{n=1}^{N_s} |y(n)|^2, \tag{2.6}$$

where $y(n)$ denotes the nth sample in a slot, and the expression for $y(n)$ is given in (2.4) and (2.5).

With a chosen threshold ϵ, the spectrum is judged occupied when $O \geq \epsilon$, otherwise the spectrum is idle. Generally, the probabilities of false alarm and miss detection can be defined as,

$$P_f(\epsilon) = \Pr(O > \epsilon | \mathcal{H}_0),$$
$$P_m(\epsilon) = \Pr(O < \epsilon | \mathcal{H}_1),$$

(2.7)

where $\mathcal{H}_0$ and $\mathcal{H}_1$ are the hypothesises when the PU is idle and busy, respectively.

Considering the difference of the received signal caused by RSI, we can achieve better sensing performance by changing the threshold according to SU$_1$'s activity. Let ϵ_0 and ϵ_1 be the thresholds when SU$_1$ is silent and busy, respectively, and now we have two sets of probabilities of false alarm and miss detection accordingly, denoted as $\{P_f^0(\epsilon_0), P_m^0(\epsilon_0)\}$ and $\{P_f^1(\epsilon_1), P_m^1(\epsilon_1)\}$, respectively.

2.2.2   Key Parameter Design

The most important constraint of the secondary networks is that their interference to the primary network must be under a certain level. In this book, we consider this constraint as the collision ratio between SUs and the PU, defined as

$$P_c = \lim_{t \to \infty} \frac{\text{Collision duration}}{\text{PU's transmission time during } [0, t]}.$$

The sensing parameters are designed according to the constraint of P_c. In the rest of this section, sensing performance is evaluated, based on which we provide the analytical design of the sensing thresholds.

Figure 2.3 shows a sketch of the design procedure of the sensing thresholds, and the derivation procedure of the secondary throughput. Specifically, the sensing error probabilities (P_m^0, P_f^0), and (P_m^1, P_f^1) are determined by thresholds ϵ_0 and ϵ_1, respectively. Combining these sensing error probabilities with the PU's arrival and departure probabilities in each slot, the state transition probabilities among State$_1$, State$_2$, State$_3$, and State$_4$ can be derived. Then we can obtain the overall miss detection and false alarm P_m and P_f, which is defined as the probability that the system stays on State$_3$ and State$_4$, respectively. Further, we consider the average time duration of collisions and unused spectrum hole, and the collision ratio and spectrum waste ratio can be given. Here, similar to the definition of collision ratio, the spectrum waste ratio is defined as

$$P_w = \lim_{t \to \infty} \frac{\text{Duration of unused spectrum hole}}{\text{Total length of spectrum holes during } [0, t]}.$$

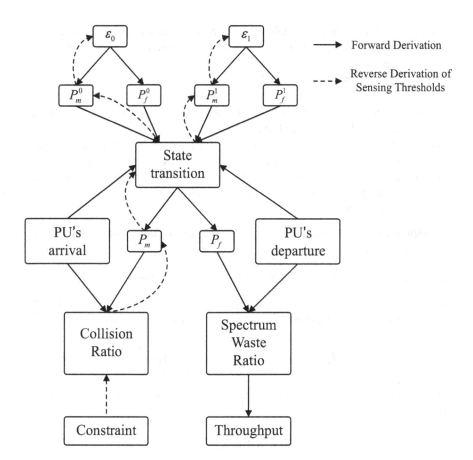

Fig. 2.3 The schematic design procedure of the sensing thresholds of the LAT protocol

Taking the collision ratio as the system constraint, the sensing thresholds can be derived reversely as shown by the dashed arrows in Fig. 2.3.

In the following of this section, we first calculate the sensing error probabilities (P_m^0, P_f^0), and (P_m^1, P_f^1), then show the state transition among the four states of the system, based on which we calculate the collision ratio as a function of sensing thresholds and PU's state change probabilities. Then, the sensing thresholds can be obtained reversely from the constraint of collision ratio.

Sensing Error Probabilities   With the statistical information of the received signal in (2.4) and (2.5), the statistical properties of O under each hypothesis can be derived. We consider the following two types of time slots:

- *Slots in which the PU changes its state:* if the PU arrives in a certain slot, the received signal power is likely to increase in the latter fraction of the slot, and the average signal power (O) is likely to be higher than the previous slots when the PU is absent. Then the probability of correct detection is higher than P_f^i, with i denotes the current activity of SU_1, i.e., $i = 0$ means that SU_1 is silent, while $i = 1$ means that SU_1 is active. Similarly, if the PU leaves in a slot, the probability of correct detection is higher than P_m^i. Note that these slots are rare in the whole traffic, we only consider the lower limits of correct detection in these slots, i.e., we set without further derivation the probabilities of correct detection to be P_f^i and P_m^i when the PU arrives or leaves, respectively.

- *Slots in which the PU remains either present or absent:* in these slots, the received signal $y(n)$ in the same slot is i.i.d., and as we assumed in Sect. 2.1.1, the number of samples N_s is sufficiently large. According to central limit theorem (CLT), the PDF of O can be approximated by a Gaussian distribution $O \sim \mathcal{N}(\mathbb{E}[|y|^2], \frac{1}{N_s}\text{var}[|y|^2])$.

Proposition 2.1. *The statistical properties and the description under each hypothesis are given in Table 2.1, in which $\gamma_s = \frac{\sigma_p \sigma_h^2}{\sigma_u^2}$ denotes the signal-to-noise ratio (SNR) in sensing, and $\gamma_i = \frac{\chi^2 \sigma_s^2}{\sigma_u^2}$ is the interference-to-noise ratio (INR).*

Proof. We first provide the general properties of the test statistics. Given that each $y(n)$ in (2.6) is i.i.d., the mean and the variance of M can be calculated as

$$\mathbb{E}[M] = \mathbb{E}\left[|y|^2\right]; \quad \text{var}[M] = \frac{1}{N_s}\text{var}\left[|y|^2\right].$$

Further, if the received signal y is complex-valued Gaussian with mean zero and variance σ_y^2, we have

$$\mathbb{E}[M] = \sigma_y^2,$$

and

$$\text{var}[M] = \frac{1}{N_s}\left(\mathbb{E}\left[|y|^4\right] - \sigma_y^4\right) = \frac{\sigma_y^4}{N_s}. \tag{2.8}$$

Then we consider the concrete form of the received signal under each hypothesis. In the LAT protocol, given the PU signal, RSI, and i.i.d. noise, the received signal y is complex-valued Gaussian with zero mean. The variance of y under the four hypothesises are as follow:

$$\sigma_y^2 = \begin{cases} (1 + \gamma_s)\sigma_u^2 & \mathcal{H}_{01}, \\ (1 + \gamma_i)\sigma_u^2 & \mathcal{H}_{10}, \\ \sigma_u^2 & \mathcal{H}_{00}, \\ (1 + \gamma_s + \gamma_i)\sigma_u^2 & \mathcal{H}_{11}. \end{cases} \tag{2.9}$$

By substituting them into (2.8), we can obtain the results in Table. 2.1.   □

Table 2.1 Properties of PDFs of LAT protocol

Hypothesis	PU	SU	$\mathbb{E}[O]$	var $[O]$
$\mathcal{H}_{00}$	Idle	Silent	σ_u^2	$\frac{\sigma_u^4}{N_s}$
$\mathcal{H}_{01}$	Busy	Silent	$(1 + \gamma_s)\sigma_u^2$	$\frac{(1+\gamma_s)^2 \sigma_u^4}{N_s}$
$\mathcal{H}_{10}$	Idle	Active	$(1 + \gamma_i)\sigma_u^2$	$\frac{(1+\gamma_i)^2 \sigma_u^4}{N_s}$
$\mathcal{H}_{11}$	Busy	Active	$(1 + \gamma_s + \gamma_i)\sigma_u^2$	$\frac{(1+\gamma_s+\gamma_i)^2 \sigma_u^4}{N_s}$

Based on Table 2.1, the sensing error probabilities can be derived.

- When SU_1 is silent and the test threshold is ϵ_0, the probability of miss detection (P_m^0) and the probability of false alarm (P_f^0) can be written as

$$P_m^0(\epsilon_0) = \Pr(O < \epsilon_0 | \mathcal{H}_{01}) = 1 - \mathcal{Q}\left(\left(\frac{\epsilon_0}{(1 + \gamma_s)\sigma_u^2} - 1\right)\sqrt{N_s}\right), \qquad (2.10)$$

and

$$P_f^0(\epsilon_0) = \Pr(O > \epsilon_0 | \mathcal{H}_{00}) = \mathcal{Q}\left(\left(\frac{\epsilon_0}{\sigma_u^2} - 1\right)\sqrt{N_s}\right), \qquad (2.11)$$

respectively, where $\mathcal{Q}(\cdot)$ is the complementary distribution function of the standard Gaussian distribution.
- Similarly, when SU_1 is transmitting with the threshold ϵ_1, the miss detection probability (P_m^1) and the false alarm probability (P_f^1) are, respectively,

$$P_m^1(\epsilon_1) = \Pr(O < \epsilon_1 | H_{11}) = 1 - \mathcal{Q}\left(\left(\frac{\epsilon_1}{(1 + \gamma_s + \gamma_i)\sigma_u^2} - 1\right)\sqrt{N_s}\right),$$
$$(2.12)$$

and

$$P_f^1(\epsilon_1) = \Pr(O > \epsilon_1 | H_{10}) = \mathcal{Q}\left(\left(\frac{\epsilon_1}{(1 + \gamma_i)\sigma_u^2} - 1\right)\sqrt{N_s}\right). \qquad (2.13)$$

State Transition and Overall Collision Probability The collision ratio is related to not only the sensing error probabilities, but also the PU's state. Thus, a joint consideration of all the four states of the system is needed. Since the slots in which the PU changes its presence are considered together with the other slots, the state transition of the system can be simplified to a discrete-time Markov chain (DTMC), in which the system can be viewed as totally time-slotted with T as the slot length. Figure 2.4 shows the state transition diagram, where we denote State$_i$ as $S_{i \mod 4}(i = 1, 2, 3, 4)$ for simplicity.

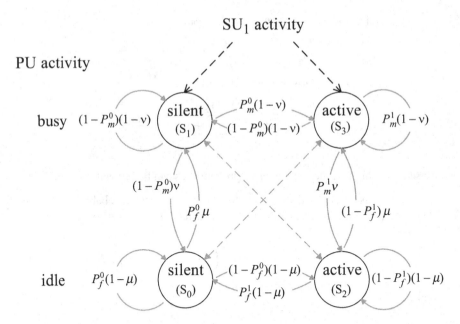

Fig. 2.4 State transition of the system

Proposition 2.2. *The probability that the system stays in the collision state S_3 is*

$$P_3 = \frac{1}{r+1} \cdot \frac{P_m^0 \left(1 - \xi\Delta\right) + \left(1 - P_f^0\right) r}{\left(1 - \xi\Delta\right)\varsigma + \xi r},$$ (2.14)

where $\xi = 1 - P_f^0 + P_f^1$, $\varsigma = 1 + P_m^0 - P_m^1$, $r = v/\mu$, and $\Delta = 1 + r - 1/\mu$.

Proof. The probability for the system staying in each state P_k $(k = 0, 1, 2, 3)$ can be calculated considering the steady-state distribution of the Markov chain:

$$\mathbf{\Psi p} = \mathbf{p},$$ (2.15)

where $\mathbf{p} = [P_0, P_1, P_2, P_3]^{\mathrm{T}}$ is the vector of steady probabilities, and $\mathbf{\Psi}$ is the state transition matrix abstracted from Fig. 2.4, which can be written as

$$\mathbf{\Psi} = \begin{bmatrix} P_f^0 \left(1 - \mu\right) & P_f^0\mu \left(1 - P_m^0\right) v & P_f^1 \left(1 - \mu\right) & \left(1 - P_m^1\right) v \\ P_f^0\mu & \left(1 - P_m^0\right)\left(1 - v\right) & P_f^1\mu & \left(1 - P_m^1\right)\left(1 - v\right) \\ \left(1 - P_f^0\right)\left(1 - \mu\right) & P_m^0 v & \left(1 - P_f^1\right)\left(1 - \mu\right) & P_m^1 v \\ \left(1 - P_f^0\right)\mu & P_m^0 \left(1 - v\right) & \left(1 - P_f^1\right)\mu & P_m^1 \left(1 - v\right) \end{bmatrix}.$$
(2.16)

Combining the constraint that $\sum\limits_{k=0}^{3} P_k = 1$, we have

$$\mathbf{p} = \frac{1}{r+1} \cdot \frac{1}{(1-\xi\Delta)\varsigma + \xi r} \cdot \begin{bmatrix} r \cdot \left(P_f^1\left(r - \varsigma\Delta\right) + 1 - P_m^1\right) \\ \left(1 - P_m^1\right)\left(1 - \xi\Delta\right) + P_f^1 r \\ r \cdot \left(\left(1 - P_f^0\right)\left(r - \varsigma\Delta\right) + P_m^0\right) \\ P_m^0\left(1 - \xi\Delta\right) + \left(1 - P_f^0\right)r \end{bmatrix} \qquad (2.17)$$

where $\xi = 1 - P_f^0 + P_f^1$, $\zeta = 1 + P_m^0 - P_m^1$, $r = v/\mu$, and $\Delta = 1 + r - 1/\mu$.

To have a check on the result, we consider the probability that the PU is busy and idle as $P_{busy} = P_1 + P_3 = \frac{\mu}{\mu+v} \approx \frac{l_1}{l_0+l_1}$ and $P_{idle} = P_0 + P_2 = \frac{v}{\mu+v} \approx \frac{l_0}{l_0+l_1}$, which are consistent with the results when we consider the PU's traffic only. □

Collision Ratio The collision of the PU and SU_1 occurs in the following two kinds of circumstances: (A) When the PU keeps occupying the spectrum and SU_1 fails to detect the presence of PU's signal in the previous slot, which is depicted in Fig. 2.4 as S_3 with the probability of P_3. The collision length is T. (B) The certain slots in which PU arrives. SU_1 is very likely to be transmitting in these slots since the PU is likely to be absent in the previous ones. The occurrence probability of this circumstance is equal to the PU's arrival rate $\frac{\mu v}{\mu+v}$.

Proposition 2.3. *The average collision length under circumstance (B), where the PU changes state can be approximated by $\frac{T}{2}$, when l_0 is large enough.*

Proof. The average collision length in this case can be calculated as

$$\bar{T}_2 = \frac{\int_0^T (T - t_0) f_0(t_0)\, dt_0}{\int_0^T f_0(t_0)\, dt_0}$$

$$= T\left(1 - l_0 - \frac{e^{-\frac{1}{l_0}}}{1 - e^{-\frac{1}{l_0}}}\right) = T \cdot \frac{\frac{1}{l_0} - 1 + e^{-\frac{1}{l_0}}}{\frac{1}{l_0}\left(1 - e^{-\frac{1}{l_0}}\right)} \qquad (2.18)$$

$$\approx T\left(\frac{1 - e^{-\frac{1}{l_0}}}{1 - e^{-\frac{1}{l_0}} + \frac{1}{l_0}e^{-\frac{1}{l_0}}}\right) \approx T\left(\frac{e^{-\frac{1}{l_0}}}{e^{-\frac{1}{l_0}} + e^{-\frac{1}{l_0}}}\right) = \frac{T}{2},$$

where the approximation is valid when l_0 is large enough. □

It is unavoidable in the LAT protocol that when the PU arrives, a short head of the signal, with the length of a SU's slot approximately, collides with the SU's signal. Combine the two circumstances, and the overall collision rate can be given by

$$P_c = \left(P_3 + \frac{1}{2} \cdot \frac{\mu\nu}{\mu + \nu}\right)/P_{busy} = \frac{\nu}{2} + \frac{P_m^0 (1 - \xi\Delta) + \left(1 - P_f^0\right) r}{(1 - \xi\Delta) \zeta + \xi r}, \qquad (2.19)$$

Design of Sensing Thresholds For the parameter design, we have a maximum allowable P_c as the system constraint, and all the parameters of the sensing process should be adjusted according to P_c. Note that Δ and r are only related to the PU's traffic, and $\{P_m^0, P_f^0\}$ and $\{P_m^1, P_f^1\}$ are closely related via thresholds ϵ_0 and ϵ_1, respectively. Thus, we actually have two independent variables of the secondary network to design to meet the constraint of P_c.

We choose P_m^0 and P_m^1 as the independent variables. With (2.19) as the only constraint, there are infinite choices of (P_m^0, P_m^1) pair. For simplicity, we set $P_m^0 = P_m^1 = P_m$, i.e., $\zeta = 1$ to reduce the degree of freedom, and the constraint can be simplified as

$$P_c = \frac{\nu}{2} + \frac{P_m (1 - \xi\Delta) + \left(1 - P_f^0\right) r}{1 + \left(\frac{1}{\mu} - 1\right)\xi}, \qquad (2.20)$$

where Δ, r, μ, and ν are only relevant to the PU traffic, and $\xi = 1 - P_f^0 + P_f^1$ can be derived from P_m via test thresholds ϵ_0 and ϵ_1.

In the rest of this part, we calculate P_m from the constraint of P_c, from which the sensing thresholds ϵ_0 and ϵ_1 can be obtained.

Combining (2.10) and (2.11), (2.12) and (2.13), we can obtain P_f^0 and P_f^1 as functions of P_m as, respectively,

$$P_f^0 (P_m) = \mathcal{Q}\left(\mathcal{Q}^{-1} (1 - P_m) (1 + \gamma_s) + \gamma_s \sqrt{N_s}\right); \qquad (2.21)$$

$$P_f^1 (P_m) = \mathcal{Q}\left(\mathcal{Q}^{-1} (1 - P_m)\left(1 + \frac{\gamma_s}{1 + \gamma_i}\right) + \frac{\gamma_s}{1 + \gamma_i} \sqrt{N_s}\right). \qquad (2.22)$$

From (2.21) and (2.22), we can find a rise of the false alarm probability when the RSI exists. This result indicates that when interference increases, the sensing performance gets worse.

With (2.21) and (2.22), ξ can be expressed as $\xi (P_m) = 1 - P_f^0 (P_m) + P_f^1 (P_m)$. With given parameters of the PU's traffic and the slot length, P_m can be solved from (2.20). Since the analytical expression of P_m is complicated, we only give some typical numerical solution in Fig. 2.5, where the sensing SNR $\gamma_s = -10$ dB, INR $\gamma_i = 5$ dB, number of samples $N_s = 200$, and r is set to be 6 to meet the real case that the typical spectrum occupancy is less than 15 % [1].

It is shown in Fig. 2.5 that when μ goes down, $P_c - \frac{\nu}{2}$ becomes a fine approximation of P_m. With the large-l_0 assumption, we regard μ as sufficiently small, and P_m is determined by

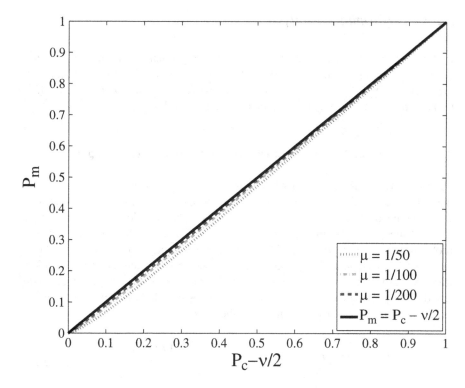

Fig. 2.5 Numerical solution of P_m; $\gamma_s = -10\,\mathrm{dB}$, $\gamma_i = 5\,\mathrm{dB}$, $N_s = 200$, and $r = 5$

$$P_m = P_c - \frac{v}{2} = P_c - \frac{1}{2}\left(1 - e^{-\frac{T}{\tau_1}}\right). \qquad (2.23)$$

This indicates that with the same constraint P_c and parameters of the PU's traffic, the required P_m gets squeezed when SU's slot length T increases.

With $P_m = P_c - v/2$, the thresholds ϵ_0 and ϵ_1 can be obtained from (2.10) and (2.12), respectively:

$$\epsilon_0 = \epsilon_0\,(P_m)|_{P_m = P_c - v/2} = \left(\frac{\mathcal{Q}^{-1}\,(1 - P_m)}{\sqrt{N_s}} + 1\right)(1 + \gamma_s)\,\sigma_u^2; \qquad (2.24)$$

$$\epsilon_1 = \epsilon_1\,(P_m)|_{P_m = P_c - v/2} = \left(\frac{\mathcal{Q}^{-1}\,(1 - P_m)}{\sqrt{N_s}} + 1\right)(1 + \gamma_s + \gamma_i)\,\sigma_u^2. \qquad (2.25)$$

A lift of sensing threshold due to the RSI (γ_i) can be found from (2.24) and (2.25), which is in accordance with the previous analysis.

2.2.3 Performance Analysis

In this section, we first evaluate the sensing performance of the LAT by the
probabilities of spectrum waste ratio under constraint of collision ratio. Then, with
the closed-form analytical secondary throughput, a tradeoff between the secondary
transmit power and throughput is elaborated theoretically.

2.2.3.1 Spectrum Utilization Efficiency and Secondary Throughput

Spectrum Waste Ratio Similar to the analysis of the collision ratio, we combine
the following two kinds of time slots to derive the spectrum waste ratio: (A) the slots
when the spectrum remains idle; and (B) the slots of the PU's departure. There exists
waste of spectrum holes in (A) when the system is in the state S_0 in Fig. 2.4, and the
probability is given by P_0 in (2.17). Every time when the SU fails to find the hole,
the waste length is T. In (B), the average waste length can be derived from the PU's
traffic with the similar method in (2.18), and it also yields $\frac{T}{2}$ of the average waste
length. The probability of the PU's departure is $\frac{\mu\nu}{\mu+\nu}$, which is same as its arrival
rate. The ratio of wasted spectrum hole is then given by

$$P_w = \left(P_0 + \frac{1}{2} \cdot \frac{\mu\nu}{\mu+\nu}\right)/P_{idle} = \frac{\mu}{2} + \frac{\left(\frac{1}{\mu}-1\right)P_f^1 + 1 - P_m}{1 + \left(\frac{1}{\mu}-1\right)\xi}. \tag{2.26}$$

Secondary Throughput SU_1's throughput can be measured with the waste ratio.
The achievable sum rate of SU_1 under perfect sensing is given as

$$R = \log_2\left(1 + \gamma_t\right), \tag{2.27}$$

where $\gamma_t = \frac{\sigma_s^2 \sigma_r^2}{\sigma_u^2}$ represents the SNR in transmission, with σ_t^2 denotes the pass loss
of the transmit channel from SU_1 to SU_2. SU_1's throughput is then written as

$$C = R \cdot (1 - P_w) = \log_2\left(1 + \gamma_t\right) \cdot \left(1 - \frac{\mu}{2} - \frac{\left(\frac{1}{\mu}-1\right)P_f^1 + 1 - P_m}{1 + \left(\frac{1}{\mu}-1\right)\xi}\right). \tag{2.28}$$

2.2.3.2 Power-Throughput Tradeoff

In the expression of SU_1's throughput in (2.28), there are two factors: R and
$(1 - P_w)$. On one hand, R is positively proportional to SU_1's transmit power σ_s^2.
On the other hand, however, the following proposition holds.

Proposition 2.4. *The spectrum waste ratio P_w increases with the secondary transmit power σ_s^2.*

Proof. Firstly, the INR γ_i increases with the transmit power and in turn lifts P_f^1, which can be seen from (2.22).

Then, we can rewrite (2.26) as

$$
\begin{aligned}
P_w &= \frac{\mu}{2} + \frac{\left(\frac{1}{\mu}-1\right)P_f^1 + 1 - P_m}{1 + \left(\frac{1}{\mu}-1\right)\xi} \\
&= \frac{\mu}{2} + \frac{\left(\frac{1}{\mu}-1\right)P_f^1 + 1 - P_m}{1 + \left(\frac{1}{\mu}-1\right)\left(1 - P_f^0\right) + \left(\frac{1}{\mu}-1\right)P_f^1} \\
&= \frac{\mu}{2} + 1 - \frac{\left(\frac{1}{\mu}-1\right)\left(1 - P_f^0\right) + P_m}{1 + \left(\frac{1}{\mu}-1\right)\left(1 - P_f^0\right) + \left(\frac{1}{\mu}-1\right)P_f^1}.
\end{aligned}
\tag{2.29}
$$

When P_f^1 increases, the third term decreases and P_w increases monotonically. Then the increase of SU_1's transmit power results in greater waste of the vacant spectrum. $\square$

Thus, there may exist a power-throughput tradeoff in this protocol: when the secondary transmit power is low, the RSI is negligible, the spectrum is used more fully with small P_w, yet the ceiling throughput is limited by R; when the transmit power increases, the sensing performance get deteriorated, while at the same time, SU_1 can transmit more data in a single slot.

Local Optimal Transmit Power[2] The analysis above indicates the existence of a mediate secondary transmit power to achieve both high spectrum utilization efficiency in time domain and high secondary throughput. To obtain this mediate value of transmit power, we differentiate the expression of the throughput to find the local optimal points of the secondary transmit power $\widehat{\sigma_s^2}$, which satisfies

$$
\left.\frac{dC}{d\sigma_s^2}\right|_{\widehat{\sigma_s^2}} = 0.
\tag{2.30}
$$

---

[2]Note that the secondary throughput is not purely convex throughout the domain of transmit power. There may exist local optimal points in low power region, while the throughput is monotonically increasing in the high power region. The point of the discussion of the power-throughput tradeoff and the calculation of the local optimal transmit power is that the secondary throughput does not monotonously increase with the transmit power, which means that SUs may not always transmit with its maximum transmit power to achieve highest throughput, instead, a mediate value may lead to better performance.

Proposition 2.5. *The local optimal power satisfies*

$$\ln\left(\gamma_t + 1\right) \cdot \frac{\exp\left(-\frac{\rho^2}{2}\right) \cdot \left(\frac{1}{\mu} - 1\right) \Xi}{\sqrt{2\pi}\,(\gamma_i + 1)^2\,\alpha} \cdot \kappa + \frac{\sigma_t^2}{\gamma_t + 1} \cdot \left(\frac{\mu}{2} - \kappa\right) = 0, \qquad (2.31)$$

where the notations are as follow:

$$\rho = \mathcal{Q}^{-1}\left(1 - P_m\right) \cdot \left(\frac{\gamma_s}{\gamma_i + 1} + 1\right) + \frac{\gamma_s}{\gamma_i + 1}\sqrt{N_s}, \text{ i.e., } \mathcal{Q}\left(\rho\right) = P_f^1,$$

$$\alpha = \left(\frac{1}{\mu} - 1\right) \cdot \left(\mathcal{Q}\left(\rho\right) - P_f^0 + 1\right) + 1,$$

$$\kappa = \frac{\left(\frac{1}{\mu} - 1\right) \cdot \left(1 - P_f^0\right) + P_m}{\left(\frac{1}{\mu} - 1\right)\left(\mathcal{Q}\left(\rho\right) - P_f^0 + 1\right) + 1},$$

$$\Xi = \gamma_s \chi^2 \left(\mathcal{Q}^{-1}\left(1 - P_m\right) + \sqrt{N_s}\right).$$

Proof. The optimal power $\widehat{\sigma_s^2}$ satisfies

$$\left. \frac{dC}{d\sigma_s^2} \right|_{\widehat{\sigma_s^2}} = 0. \qquad (2.32)$$

The differentiation of the secondary throughput can be derived as

$$\frac{dC}{d\sigma_s^2} = -\log_2\left(\gamma_t + 1\right) \cdot \frac{\exp\left(-\frac{\rho^2}{2}\right) \cdot \left(\frac{1}{\mu} - 1\right) \cdot \frac{\gamma_s \chi^2 \left(\mathcal{Q}^{-1}(1 - P_m) + \sqrt{N_s}\right)}{(\gamma_i + 1)^2}}{\sqrt{2\pi}\left[\left(\frac{1}{\mu} - 1\right) \cdot \left(\mathcal{Q}\left(\rho\right) - P_f^0 + 1\right) + 1\right]^2}$$

$$\cdot \left(\left(\frac{1}{\mu} - 1\right) \cdot \left(1 - P_f^0\right) + P_m\right)$$

$$- \frac{\sigma_t^2}{\ln 2 \cdot \left(\gamma_t + 1\right)} \cdot \left(\frac{\mu}{2} + \frac{\mathcal{Q}\left(\rho\right) \cdot \left(\frac{1}{\mu} - 1\right) - P_m + 1}{\left(\frac{1}{\mu} - 1\right) \cdot \left(\mathcal{Q}\left(\rho\right) - P_f^0 + 1\right) + 1} - 1\right), \qquad (2.33)$$

where $\rho = \mathcal{Q}^{-1}\left(1 - P_m\right) \cdot \left(\frac{\gamma_s}{\gamma_i + 1} + 1\right) + \frac{\gamma_s}{\gamma_i + 1}\sqrt{N_s}$, i.e., $\mathcal{Q}\left(\rho\right) = P_f^1$. With $\frac{dC}{d\sigma_s^2} = 0$, we have

$$\ln(\gamma_t + 1) \cdot \frac{\exp\left(-\frac{\rho^2}{2}\right) \cdot \left(\frac{1}{\mu} - 1\right) \cdot \frac{\gamma_s \chi^2 \left(Q^{-1}(1-P_m) + \sqrt{N_s}\right)}{(\gamma_i + 1)^2}}{\sqrt{2\pi}\left[\left(\frac{1}{\mu} - 1\right)\left(Q(\rho) - P_f^0 + 1\right) + 1\right]^2} \cdot \left(\left(\frac{1}{\mu} - 1\right) \cdot \left(1 - P_f^0\right) + P_m\right)$$

$$+ \frac{\sigma_t^2}{\gamma_t + 1} \cdot \left(\frac{\mu}{2} + \frac{\left(\frac{1}{\mu} - 1\right) \cdot Q(\rho) - P_m + 1}{\left(\frac{1}{\mu} - 1\right)\left(Q(\rho) - P_f^0 + 1\right) + 1} - 1\right) = 0, \tag{2.34}$$

and therefore,

$$\ln(\gamma_t + 1) \cdot \frac{\exp\left(-\frac{\rho^2}{2}\right) \cdot \left(\frac{1}{\mu} - 1\right) \Xi}{\sqrt{2\pi}\,(\gamma_i + 1)^2 \alpha} \cdot \kappa + \frac{\sigma_t^2}{\gamma_t + 1} \cdot \left(\frac{\mu}{2} - \kappa\right) = 0. \tag{2.35}$$

When μ is sufficiently small, the notations can be simplified as

$$\alpha = \frac{1}{\mu} \cdot \left(Q(\rho) - P_f^0 + 1\right),$$

$$\kappa = \frac{1 - P_f^0}{Q(\rho) - P_f^0 + 1}, \tag{2.36}$$

and (2.35) becomes

$$\frac{\ln(\gamma_t + 1)}{(\gamma_i + 1)^2} \cdot \frac{\exp\left(-\frac{\rho^2}{2}\right) \cdot \Xi}{\sqrt{2\pi}\left(Q(\rho) - P_f^0 + 1\right)} \cdot \kappa - \frac{\sigma_t^2 \kappa}{\gamma_t + 1} = 0, \tag{2.37}$$

i.e.,

$$\exp\left(-\frac{\rho^2}{2}\right) \frac{(\gamma_t + 1)\ln(\gamma_t + 1)}{(\gamma_i + 1)^2} = \frac{\sqrt{2\pi}\sigma_t^2}{\Xi}\left(1 - P_f^0 + Q(\rho)\right). \tag{2.38}$$

$\square$

In (2.31), with σ_s^2 as the only unknown variable, it can be calculated numerically. To obtain better comprehension about the properties of the local optimal transmit power, we consider the case when μ is sufficiently small, and (2.31) can be simplified as

$$\exp\left(-\frac{\rho^2}{2}\right) \frac{(\gamma_t + 1)\ln(\gamma_t + 1)}{(\gamma_i + 1)^2} = \frac{\sqrt{2\pi}\sigma_t^2}{\Xi}\left(1 - P_f^0 + Q(\rho)\right). \tag{2.39}$$

Existing Conditions of the Local Optimal Transmit Power The left side of (2.39) is a convex curve of σ_s^2 with a single maximum. When σ_s^2 goes to zero or infinity, the value of the left side goes to zero. The value of the right side changes from $\frac{\sqrt{2\pi}\sigma_t^2}{\Xi}$ to $\frac{\sqrt{2\pi}\sigma_t^2}{\Xi}\left(2 - P_f^0 - P_m\right)$. We can roughly say that when the maximum of the left side is larger than either $\frac{\sqrt{2\pi}\sigma_t^2}{\Xi}$ or $\frac{\sqrt{2\pi}\sigma_t^2}{\Xi}\left(2 - P_f^0 - P_m\right)$, there would be two solutions to (2.39). When the maximum of the left side is smaller than the minimum of the right, on the other hand, no solution exists.

- When there is no solution, the curves of transmit power on the left and right sides never meet. Since the right side of (2.39) is always far above zero and the left can go to zero when the transmit power is extremely high or low, we can safely say that the left side is always smaller than the right, i.e.,

$$\exp\left(-\frac{\rho^2}{2}\right)\frac{(\gamma_t + 1)\ln(\gamma_t + 1)}{(\gamma_i + 1)^2} < \frac{\sqrt{2\pi}\sigma_t^2}{\Xi}\left(1 - P_f^0 + \mathcal{Q}(\rho)\right). \qquad (2.40)$$

Substituting the inequation to (2.33), we have $\frac{dC}{d\sigma_s^2} > 0$, which indicates that the secondary throughput would increase with the transmit power monotonously.
- When the solutions of (2.39) exist, we discuss the sign of $\frac{dC}{d\sigma_s^2}$ piecewise. When the power is low or high enough, the left side is small, while the right remains considerable. The solid red curve (maximum of the left side of (2.39)) is below the dash-dotted blue one (value of the right side of (2.39)), and $\frac{dC}{d\sigma_s^2} > 0$. When the power is between the two solutions, we have $\frac{dC}{d\sigma_s^2} < 0$. Thus, at the smaller solution, $\frac{d^2C}{d(\sigma_s^2)^2} < 0$, and this is the local optimal transmit power $\widehat{\sigma_s^2}$ to achieve local maximum throughput. Similarly, the larger solution denotes the local minimum of the throughput.

The maximum of the left side and the corresponding value of the right side is shown in Fig. 2.6, in which the parameters are listed in Table 2.3. It is shown that when χ^2 is smaller than 0.8, the maximum of the left is larger than the corresponding value of the right, and (2.39) will have solutions and power-throughput is likely to exist. When χ^2 is greater than 0.85, there may be no tradeoff between the transmit power and secondary throughput.

2.2.3.3 Comparison with the Listen-Before-Talk Protocol

To better understand the LAT protocol, we compare it with the conventional LBT in this section. There exist limitations for both LBT and LAT protocols. In the LBT, the data transmission time is reduced because of spectrum sensing, and the overall throughput is also affected by spatial correlation. In the LAT, RSI is the main problem that decreases the performance. In this section, we first briefly introduce the LBT protocol and derive its sensing performance and throughput. Then comparison is made base on the analytical results and a switching scheme between the LAT and LBT is proposed to maximize SU's throughput.

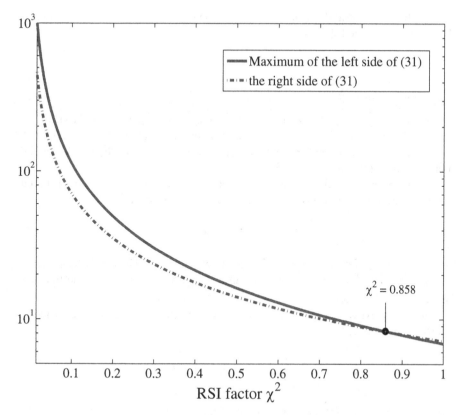

Fig. 2.6 Numerical verification of the existence of the power-throughput tradeoff

The Listen-Before-Talk Protocol[3] In the conventional CRNs, every SU's slot is divided into two sub-slots: the sensing sub-slot and the transmission sub-slot. In the sensing sub-slot with duration T_s, SU_1 uses both Ant_1 and Ant_2 for sensing, and in the transmission sub-slot, both antennas transmit data to the two antennas at SU_2 as a 2×2 MIMO system. In the LBT protocol, sensing process is isolated from secondary data transmission. Thus, transmit power is no longer a harmful factor to the sensing performance, and theoretically the secondary throughput always increases with the transmit power. At the same time, considerations of some more factors are needed only in the LBT protocol: the spatial correlation among the antennas and the ratio of sensing time T_s over the whole slot length T.

In the sensing process, without the influence of self-interference, there are only two hypothesises to test. Let $\mathcal{H}_0$ be the hypothesis under which the spectrum is idle,

---

[3]Note that for comparison fairness, we use the same system model shown in Fig. 2.1 for both the LAT and LBT protocols. Thus, for the LBT protocol, each SU is equipped with two antennas, and spatial diversity needs to be considered.

and $\mathcal{H}_1$ be the one that the spectrum is busy. The received signal at SU_1 under each hypothesis takes the following form:

$$y = \begin{cases} h_s s_p + u, & \mathcal{H}_1, \\ u, & \mathcal{H}_0, \end{cases} \tag{2.41}$$

where s_p is the PU's signal, h_s is a 2×1 vector denoting the channel from PU to SU_1, and u is the noise vector. With the separable correlation model[12], h_s can be expressed as $h_s = \Phi_s^{1/2} h_{s0}$ where $h_{s0} \sim \mathcal{CN}(0, \sigma_h^2)$, and Φ_s stands for the normalized correlation matrix at SU_1.

When a spectrum hole is detected, the secondary data is transmitted in a spatial multiplexing way. The received signal at SU_2 can be written as

$$r = H_t s_t + u, \tag{2.42}$$

where $H_t = \Phi_r^{1/2} H_{t0} \Phi_t^{1/2}$ is a 2×2 channel matrix from SU_1 to SU_2, H_{t0} is i.i.d. complex-valued Gaussian with zero mean and variance σ_h^2, and s_t is the transmit signal vector with variance σ_s^2.

We use the exponential correlation model[13] to determine the spatial correlation matrixes Φ_s, Φ_r, and Φ_t as

$$\Phi_\Lambda = \begin{pmatrix} 1 & \beta_\Lambda \\ \beta_\Lambda^* & 1 \end{pmatrix}, \quad |\beta_\Lambda| \in [0,1), \quad \Lambda = \{s, t, r\}, \tag{2.43}$$

where the factor β_Λ is the spatial correlation factor.

Following the same procedure of the derivation in the LAT, the theoretical ratio of spectrum waste P'_w under the same constraint of collision percentage P_c of the LAT can be obtained:

Proposition 2.6. *The ratio of spectrum waste P'_w in the LBT protocol can be expressed as*

$$P'_w = \lambda + \frac{1 - \lambda^2}{2} \mu + (1 - \lambda) \mathcal{Q} \left(\mathcal{Q}^{-1} \left(1 - \frac{P_c}{1 - \lambda} + \frac{1 + \lambda}{2} v \right) \eta + \gamma_s \sqrt{2 \lambda N_s} \right), \tag{2.44}$$

where $\lambda = T_s/T$ is the sensing ratio in a whole slot, and the parameter $\eta :=$ $\left((\beta_s \gamma_s)^2 + (\gamma_s + 1)^2 \right)^{1/2}$.

Proof. Same as the sensing process in the LAT protocol, we take energy detection as the sensing scheme. The test statics M' can be written as

$$M' = \frac{1}{N'_s} \sum_{n=1}^{N'_s} \frac{|y_1(n)|^2 + |y_2(n)|^2}{2}, \tag{2.45}$$

Table 2.2 Statistical
properties of M'

	Mean ($\mathbb{E}[M']$)	Variance (var $[M']$)
$\mathcal{H}_0$	σ_u^2	$\frac{\sigma_u^4}{2N_s'}$
$\mathcal{H}_1$	$(\gamma_s + 1)\sigma_u^2$	$\frac{\left[(\beta_s \gamma_s)^2 + (\gamma_s + 1)^2\right]\sigma_u^4}{2N_s'}$

where $N_s' = f_s T_s = \lambda N_s$ is the number of samples in each sensing sub-slot, and y is specified in (2.41). The distribution of M' can be approximated by a Gaussian distribution according to CLT, given that each sample $\frac{|y_1|^2 + |y_2|^2}{2}$ is i.i.d. and N_s' is sufficiently large.

The statistical properties of M' under both hypothesises are presented in Table 2.2, in which $\mathbb{E}[M']$ and var$[M']$ under $\mathcal{H}_0$ can be obtained following the same procedure in the proof of Proposition 2.1, and those under $\mathcal{H}_1$ can be derived as follow.

$$\mathbb{E}[M'] = \mathbb{E}\left[\frac{|y_1|^2 + |y_2|^2}{2}\right] = \frac{1}{2}\mathbb{E}[\mathbf{y}^H \mathbf{y}]$$

$$= \frac{1}{2}\mathbb{E}\left[\mathbf{h}_{s0}^H (\mathbf{\Phi}_s^{1/2})^H \mathbf{\Phi}_s^{1/2} \mathbf{h}_{s0} + 2\sigma_u^2\right] = (\gamma_s + 1)\sigma_u^2,$$

and

$$\text{var}[M'] = \frac{1}{4N_s'}\left(\mathbb{E}[\mathbf{y}^H \mathbf{y}]^2 - 4(\gamma_s + 1)^2 \sigma_u^4\right).$$

Suppose

$$\mathbf{\Phi}^{1/2} = \begin{bmatrix} a & b \\ b* & a* \end{bmatrix}, \text{ and } h_{s0} = \begin{pmatrix} h_1 \\ h_2 \end{pmatrix}, \tag{2.46}$$

where a and b satisfy

$$\begin{cases} |a|^2 + |b|^2 = 1, \\ 2\Re[ab] = \beta. \end{cases} \tag{2.47}$$

Then

$$\mathbb{E}[\mathbf{y}^H \mathbf{y}]^2 = \mathbb{E}\left[\left((|ah_1 + bh_2 + u_1|^2 + |bh_1 + ah_2 + u_2|^2)\right)^2\right]$$

$$= \mathbb{E}\left[|ah_1 + bh_2 + u_1|^4 + |bh_1 + ah_2 + u_2|^4 + 2|ah_1 + bh_2 + u_1|^2 |bh_1 + ah_2 + u_2|^2\right]$$

$$= 2(\gamma_s + 1)^2 + 2(\gamma_s + 1)^2 + (1 + \beta_s^2)\gamma_s^2 + 1 + 2\gamma_s$$

$$= 4(\gamma_s + 1)^2 + \beta_s^2 \gamma_s^2 + (\gamma_s + 1)^2.$$

Thus,

$$\text{var}\left[M'\right] = \frac{1}{2N'_s}\left(\left[(\beta_s\gamma_s)^2 + (\gamma_s + 1)^2\right]\sigma_u^4\right) = \frac{\eta^2\sigma_u^4}{2N'_s}.$$

With a given threshold ϵ, the probability of false alarm and miss detection can be derived as, respectively,

$$P'_f(\epsilon; T_s) = Q\left(\left(\frac{\epsilon}{\sigma_u^2} - 1\right)\sqrt{2N'_s}\right), \tag{2.48}$$

$$P'_m(\epsilon; T_s) = 1 - Q\left(\frac{\epsilon - (\gamma_s + 1)\sigma_u^2}{\eta\sigma_u^2}\sqrt{2N'_s}\right). \tag{2.49}$$

The collision in the LBT protocol also includes two parts: collision caused by miss detection and by the PU's arrival. However, since SU_1 never transmits in the first part of each slot, the derivation of collision rate is slightly different from that in the LAT. When SU_1 fails to detect the presentence of the PU's signal, the collision length is $(1 - \lambda)T$, and the probability for this case is P'_m. In the slots that the PU arrives, the average collision length is given by

$$
\begin{aligned}
\bar{T}'_2 &= \frac{\displaystyle\int_{mT}^{mT+T_s}(T - T_s)f_0(t_0)\,dt_0 + \int_{mT+T_s}^{(m+1)T}((m+1)T - t_0)f(t_0)\,dt_0}{\displaystyle\int_{mT}^{(m+1)T}f_0(t_0)\,dt_0} \\[2mm]
&= T \cdot \frac{1 - \lambda - l_0\left(e^{-\frac{\lambda}{l_0}} - e^{-\frac{1}{l_0}}\right)}{1 - e^{-\frac{1}{l_0}}} \approx \frac{1 - \lambda^2}{2} \cdot T.
\end{aligned}
\tag{2.50}
$$

Then the collision rate can be derived as

$$P'_c = (1 - \lambda)P'_m + \frac{(1 - \lambda^2)\,v}{2}. \tag{2.51}$$

Under the same constraint of collision rate P_c, the maximum allowable miss detection probability is

$$P'_m = \frac{P_c}{1 - \lambda} - \frac{1 + \lambda}{2}v, \tag{2.52}$$

and the analytical false alarm probability can be derived from (2.48) and (2.49) as

$$P'_f(P'_m; \lambda) = Q\left(Q^{-1}(1 - P'_m)\eta + \gamma_s\sqrt{2\lambda N_s}\right). \tag{2.53}$$

$\square$

In transmission, we set the same average transmit power in every time slot of the two protocol for comparison fairness. Thus, the transmit power at each antenna is

$$\sigma_{each}^2 = \frac{\sigma_s^2}{2} \cdot \frac{1}{1-\lambda}, \tag{2.54}$$

and the average sum rate is given by

$$R' = \mathbb{E}\left[\log_2 \det\left(\mathbf{I} + \frac{\sigma_{each}}{\sigma_u^2}\mathbf{H}_t\mathbf{H}_t^H\right)\right] = \mathbb{E}\left[\log_2 \det\left(\mathbf{I} + \frac{\gamma_t}{2(1-\lambda)}\mathbf{H}_t\mathbf{H}_t^H\right)\right], \tag{2.55}$$

and the throughput is expressed as

$$C' = R' \cdot \left(1 - P_w'\right)$$
$$= R' \cdot (1-\lambda)\left(1 - \mathcal{Q}\left(\mathcal{Q}^{-1}\left(\left(1 - \frac{P_c}{1-\lambda} + \frac{1+\lambda}{2}v\right)\right)\eta + \gamma_s\sqrt{2\lambda N_s}\right) - \frac{1+\lambda}{2}\mu\right). \tag{2.56}$$

Comparison and Switching Between the LAT and LBT Protocols The expressions of the throughput of the LAT and the LBT in (2.28) and (2.56) show that the performance of the two protocols are influenced by different parameters. In the LBT protocol, sensing duration over the whole slot λ is an important parameter, which is closely studied and optimized in [6], while in the LAT, SUs do not need to sacrifice any time for sensing. Also, the spatial correlation coefficient β_Λ is unique in the LBT. Both sensing and transmission performance in the LBT protocol deteriorate with the increase of the spatial correlation. In the LAT protocol, with the sensing process largely influenced by self-interference, the RSI factor χ holds great importance. In the LBT protocol, however, the self-interference is without consideration.

Thus, with certain conditions of the environment such as given γ_s, χ, β, etc., we can choose one protocol over the two to achieve higher secondary throughput. The switching criterion can be derived based on (2.28) and (2.56) as

$$\text{Adaptive Switching} = \begin{cases} \text{LBT,} & C' - C \geq 0, \\ \text{LAT,} & C' - C < 0, \end{cases} \tag{2.57}$$

where the optimal switching point can be easily obtained by solving $C' = C$.

Note that from (2.57), it implies that the switching point is related to the following statistical factors: SNR (γ_s, γ_t) and transmit power (P_t) during sensing and data transmission, spatial correlation coefficients (β_s, β_t, β_r) in the LBT protocol, and RSI factor (χ) in the LAT protocol. We give the qualitative analysis about how the values of the above parameters influence the selection between the two protocols.

- Spatial correlation (β_Λ) and RSI factor (χ): these parameters affect only one of the two protocols, and the influence on the throughput is monotonic. According to (2.44), the spectrum waste ratio of the LBT rises with η, which, according to its definition, is positively related to the spatial correlation coefficient of the sensing channel β_s. Also, we can observe from (2.55) that the sum rate of the LBT R' decreases with both the coefficients of the transmit and receive channels β_t and β_r. Thus, with the increase of β_Λ the two factors of the throughput in the LBT C' in (2.56) R' and $1 - P'_w$ both degrade, and C' decreases. When β_Λ goes to 1, the advantage of MIMO degrades and the transmit time remains limited. The performance of the LBT is much worse than the LAT. Similarly, in the LAT, the spectrum waste ratio P_w rises with the RSI factor χ via the increase of γ_i and in turn P_f^1, which is shown in (2.22) and (2.26). Thus, if χ cannot be suppressed to a low level, RSI deteriorates the sensing performance in the LAT protocol significantly, and thus, the LBT outperforms the LAT protocol.
- Sensing SNR (γ_s): γ_s influences the sensing performance of both protocols. Note that the main advantage of the LAT is longer sensing and transmission time, while the LBT benefits from spatial multiplexing. When γ_s is small, it is likely that having more sensing samples takes great advantage, which indicates the suitability of the LAT. When γ_s is large and SUs under both protocols can clearly detect the existence of the PU, the LBT may become a better choice.
- Secondary transmit power (σ_s^2): for the LAT protocol, the RSI, which is key factor that deteriorates sensing performance, is proportional to σ_s^2. As analyzed in Sect. 2.2.3.2, on one hand, the sensing performance degrades with the increase RSI, i.e., the sensing degrades with the increase of the transmit power; on the other hand, the achievable rate of secondary users increases with the transmit power. Thus, the secondary throughput in the LAT protocol may be non-monotonous with the transmit power, a power-throughput tradeoff may exist. Meanwhile, in the LBT protocol, since the sensing is independent with transmission, larger transmit power directly leads to higher throughput. These features indicates that in the high-power region, the LBT may be a better option.

2.2.3.4 Results

In this part, simulation results are presented to evaluate the performance of the proposed LAT protocol. Table 2.3 lists some default parameters in the simulation. For simplification, we set the spatial correlation coefficients $\beta_s = \beta_r = \beta_t = \beta$.

Power-Throughput Relationship of the LAT Protocol As is shown in Fig. 2.7, we consider the throughput performance of the LAT protocol in terms of secondary transmit power. The solid and dotted lines represent the analytical performance of the LAT protocol, and the asterisks (*) denote the analytical local optimal transmit power. The small circles are the simulated results, which match the analytical performance well. The thin solid line depicts the ideal case with perfect RSI cancelation. Without RSI, the sensing performance is no longer affected by transmit

Table 2.3 Simulation parameters

Parameters	Value
The number of samples a whole time-slot (N_s)	300
The probabilities of the PU's arrival in the next slot (μ)	1/500
The probabilities of the PU's departure in the next slot (v)	6/500
Normalized transmit power (σ_s^2/σ_u^2)	10 dB
SNR in sensing process (γ_s)	−5 dB
RSI factor (χ)	0.4
The spatial correlation coefficient (β)	0.8
Probability of collision (P_c)	0.1

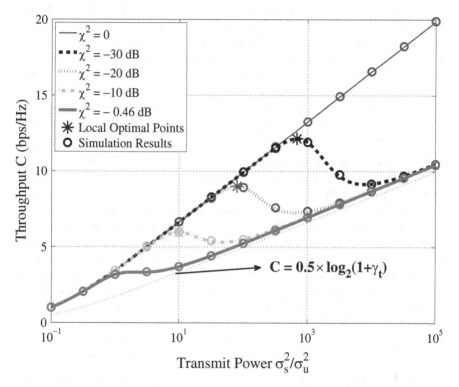

Fig. 2.7 Power-throughput curve in terms of different RSI factor χ^2, where the probability of the PU's arrival $\mu = 1/500$, departure $v = 6/500$, the collision ratio $P_c = 0.1$, the sample number of a slot N_s is 300, sensing SNR $\gamma_s = -5$ dB

power, and the throughput always goes up with the power. This line is also the upperbound of the LAT performance. The thick dash-dotted, dotted and dash lines in the middle are the typical cases, in which we can clearly observe the power-throughput tradeoff and identify the local optimal power, which is calculated from (2.39). With the decrease of RSI (χ^2 from 0.1 to 0.01 to 0.001), the local optimal transmit power increases, and the corresponding throughput goes to a higher level. This makes sense since the smaller the RSI is, the better it approaches the ideal case, and the deterioration cause by self-interference becomes dominant under a higher power. According to Fig. 2.6, when χ^2 is sufficiently large (0.85 in the figure), there exists no power-throughput tradeoff. We verify this result by the thick solid line denoting the cases when $\chi^2 = 0.9$. No local optimal point can be found in this curve, and the numerical results show that the differentiation is always positive.

One noticeable feature of Fig. 2.7 is that when self-interference exists, all curves approach the thin dotted line $C = 0.5 \log_2 (1 + \gamma_t)$ when the power goes up. This line indicates the case that the spectrum waste is 0.5. When the transmit power is too large, severe self-interference largely degrades the performance of spectrum sensing, and the false alarm probability becomes unbearably high. It is likely that whenever SU_1 begins transmission, the spectrum sensing result falsely indicates that the PU has arrived due to false alarm, and SU_1 stops transmission in the next slot. Once SU_1 becomes silent, it can clearly detect the PU's absence, and begins transmission in the next slot again. And the state of SU_1 changes every slot even when the PU does not arrive at all. In this case, the utility efficiency of the spectrum hole is approximately 0.5, which is clearly shown in Fig. 2.7. Also, it can be seen that the larger χ^2 is, the earlier the sensing gets unbearable and the throughput approaches the orange line.

Sensing Performance In this section, we use the receiver operating characteristic curves (ROCs) to present the sensing performance. In Fig. 2.8, with the sensing SNR γ_s fixed on -8 dB, and the spatial correlation $\beta = 0.85$, we have the relation between the collision ratio and spectrum waste ratio. In Fig. 2.8, smaller area under a curve denotes better sensing performance, i.e., the sensing performance of the LAT is much better than the LBT. It can be seen that for the curves in the LAT protocol, the solid red line is lower than the dashed pink one, which indicates that smaller RSI leads to better sensing performance. It is noteworthy that the ratio of spectrum waste of the LBT protocol can never be lower than λ, while that of the LAT can be quite close to zero if the PU's state change is sufficiently slow. Also, in the left side of Fig. 2.8 where the allowable P_c is small, P_w in the LAT decreases much more sharply with the increase of P_c than P'_w in the LBT, which also implies better sensing performance of the LAT in the real case when P_c is strictly constrained.

Impact of the RSI Factor χ^2 In Fig. 2.9, we consider the impact of the RSI factor on the sensing performance. We fix the constraint of P_c as 0.1, and evaluate the spectrum waste ratio under various χ^2. It can be seen from Fig. 2.9 that with the increase of χ^2, the spectrum waste ratio increases from zero to approximately 0.5. This is reasonable in the sense that with sufficiently small RSI factor, the RSI can be neglect compared with PU's signal and noise, and SUs can fully utilize the spectrum holes. When the RSI factor is moderate or close to 1, which indicates that the RSI cannot be suppressed well, the secondary signal may overwhelm the PU's signal, leading to unreliability of sensing, and the SUs are likely to stop communication due to false alarm. Note that the asymptotic value of the spectrum waste ratio when the RSI is large is 0.5, which is in accordance with the results in Fig. 2.7.

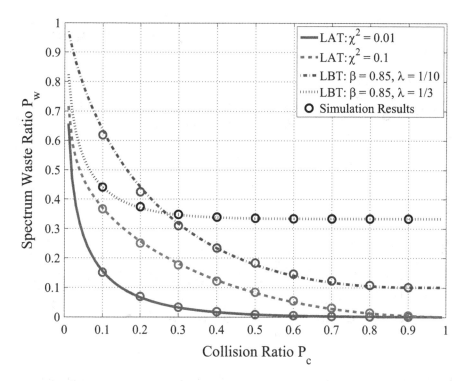

Fig. 2.8 ROCs in sensing. In this figure, the probability of the PU's arrival $\mu = 1/500$, departure $\nu = 6/500$, the sample number of a slot N_s is 300, normalized secondary transmit power $\sigma_s^2/\sigma_u^2 = 10$ dB, sensing SNR $\gamma_s = -8$ dB, the RSI factor χ^2 in the LAT varies between 0.1 and 0.01, the spatial correlation coefficient β is 0.85, and the ratio of sensing duration changes between 1/3 and 1/10

Besides, it can be seen that when the normalized power of RSI ($\chi^2\sigma_s^2/\sigma_u^2$) ranges from approximately [0.1 ,10], the spectrum waste ratio changes fast, and when the normalized power of RSI is below 0.1, the waste ratio remains at a low level. This feature can be utilized to design the protocol parameters to achieve full utilization of the spectrum holes.

Switch Scheme by Different Sensing SNR γ_s In Fig. 2.10, we consider the comparison and switching point based on the SNR in sensing (γ_s) under different spatial correlation coefficient and RSI factor. The probability of collision P_c is fixed on 0.1. We investigate the cases when the SIS factor χ^2 is 0.01 and 0.1, and when the sensing ratio in the LBT protocol changes between $\frac{1}{3}$ and $\frac{1}{10}$. It can be shown that when γ_s is low, the proposed LAT protocol can achieve better performance since the sensing time is long enough. As the sensing SNR increases, the sensing performance of the LBT protocol improves and the advantage of spatial multiplexing can be observed, and thus, the LBT outperforms the LAT if the sensing duration is carefully designed.

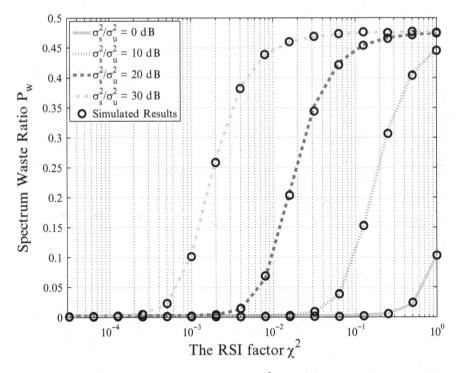

Fig. 2.9 Secondary throughput versus the RSI factor χ^2, in which the probability of the PU's arrival $\mu = 1/500$, departure $\nu = 6/500$, the collision ratio is 0.1, the sample number of a slot N_s is 300, sensing SNR $\gamma_s = -5\,\mathrm{dB}$, and the normalized secondary transmit power σ_s^2/σ_u^2 varies from 0 to 30 dB

2.3 Conclusion

In this chapter, we first proposed a LAT protocol for the simplest model of FD-CRNs that allows SUs to simultaneously sense and access the spectrum holes. We analyzed the performance of the LAT protocol and studied the tradeoff between the secondary transmit power and SU's throughput. We found that the increment of transmit power does not always yield the improvement of SU's throughput, and a mediate value is required to achieve the optimal performance. The theoretical optimal transmit power and its existing condition have been derived, and verified through simulation results. Besides, a switching scheme between the LAT and the conventional LBT protocol was provided to further improve the throughput of SUs. Specifically, when sensing SNR is low, the RSI is small, yet the spatial correlation is severe, the LAT protocol outperforms the LBT due to its better sensing performance. Otherwise, the LBT protocol may be a better option.

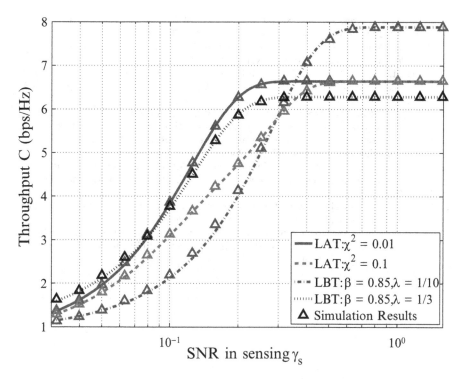

Fig. 2.10 Secondary throughput versus sensing SNR γ_s, in which the probability of the PU's arrival $\mu = 1/500$, departure $\nu = 6/500$, the collision ratio is 0.1, the sample number of a slot N_s is 300, normalized secondary transmit power $\sigma_s^2/\sigma_u^2 = 10$ dB, the RSI factor χ^2 in the LAT varies between 0.1 and 0.01, the spatial correlation coefficient β is 0.85, and the ratio of sensing duration changes between 1/3 and 1/10

References

1. Federal Communications Commission, "Spectrum Policy Task Force", Rep. ET Docket no. 02–135, Nov. 2002.
2. J. Mitola and G. Q. Maguire, "Cognitive Radio: Making Software Radios more Personal," *Personal Communications, IEEE*, vol. 6, no. 4, pp. 13–18, Aug. 1999.
3. J. Mitola, "Cognitive Radio—An Integrated Agent Architecture for Software Defined Radio," Ph.D. Thesis, Royal Institute of Technology, Sweden, May. 2000.
4. I. F. Akyildiz, W. Y. Lee, M. C. Vuran, and S.Mohanty, "Next Generation/Dynamic Spectrum Access/Cognitive Radio Wireless Networks: A Survey," *Computer Networks*, vol. 50, no. 13, pp. 2127–2159, Sep. 2006.
5. T. Yucek and H. Arslan, "A Survey of Spectrum Sensing Algorithms for Cognitive Radio Applications," *IEEE Communications Surveys & Tutorials*, vol. 11, no. 1, pp. 116–130, Mar. 2009.
6. Y. C. Liang, Y. Zeng, E. C. Y. Peh, and A. T. Hoang, "Sensing-Throughput Tradeoff for Cognitive Nadio Networks," *IEEE Trans. on Wireless Comm.*, vol. 7, no. 4, pp. 1326–1337, Apr. 2008.

7. S. Huang, X. Liu, and Z. Ding, "Short Paper: On Optimal Sensing and Transmission Strategies for Dynamic Spectrum Access," in *Proc. IEEE DySPAN*, Chicago, IL, Oct. 2008.
8. Y. Liao, T. Wang, L. Song, and Z. Han, "Listen-and-Talk: Full-Duplex Cognitive Radio," in *IEEE Proc. Globecom'2014*, Austin, TX, Dec. 2014.
9. Y. Liao, L. Song, Z. Han, and Y. Li, "Full-Duplex Cognitive Radio: A New Design Paradigm for Enhancing Spectrum Usage," *IEEE Communications Magazine*, vol. 53, no. 5, pp. 138–145, May 2015.
10. E. Everett, A. Sahai, and A. Sabharwal, "Passive Self-Interference Suppression for Full-Duplex Infrastructure Nodes," *IEEE Trans. on Wireless Comm.*, vol. 13, no. 2, pp. 680–694, Feb. 2014.
11. S. Huang, X. Liu, and Z. Ding, "Opportunistic Spectrum Access in Cognitive Radio Networks," in *IEEE InfoCom 2008*, Phoenix, AZ, Apr. 2008.
12. M. Kiessling and J. Speidel, "Mutual Information of MIMO Channels in Correlated Rayleigh Fading Environments - a General Solution," in *IEEE ICC*, vol. 2, pp. 814–818, Paris, France, Jun. 2004.
13. S. L. Loyka, "Channel Capacity of MIMO Architecture Using the Exponential Correlation Matrix," *IEEE Comm. Lett.*, vol. 5, no. 9, pp. 369–371, Sep. 2001.
14. S. Sengupta, S. Brahma, M. Chatterjee, and S. Shankar, "Enhancements to Cognitive Radio Based IEEE 802.22 Air-interface," in *Proc. IEEE International Conference on Communications (ICC)*, Glasgow, Scotland, Jun. 2007.

Chapter 3
Extensions of the LAT Protocol

The basic LAT protocol considers only one PU and one pair of SUs, which is quite limited. In this chapter, we consider the scenarios with multiple FD users that adopt the LAT protocol as their basic PHY sensing protocol. Firstly, we study the cooperative spectrum sensing under the LAT protocol, in which the interference among cooperative users makes the cooperation different from conventional cooperative schemes [1]. We provide a feasible cooperation scheme that is suitable for FD SUs in Sect. 3.1. Then, we consider both *distributed* [2] and *centralized* network [3] scenarios that require design of MAC layer protocols for harmonious multiple access and resource allocation. In distributed scenarios, multiple FD SUs contend for the same spectrum resources without any central controller. Thus, an effective multiple access scheme that can not only mitigate collision among SUs, but also fully explore the benefits of FD is needed. In centralized network, a FD cognitive AP, which can be regarded as a central controller over all SUs beneath, needs to sense the spectrum opportunities and allocate the temporal, frequency and power resources to the SUs properly, so that SUs can enjoy high data rate and the sensing performance of the AP can be guaranteed.

3.1 Extension 1: Cooperative Spectrum Sensing

In this section, we extend the simple model shown in Fig. 2.1 to the scenario with multiple cooperative SUs to further improve sensing performance.

As shown in Fig. 3.1, we consider a CRN consisting of one PU, one fusion center (FC), and M SUs denoted by $SU_1, SU_2, \ldots, SU_M$, each of which equips two antennas Ant_{i1} and Ant_{i2} for $i = 1, \ldots, M$. In each time slot with duration T, the SUs sense the spectrum with local energy detectors, and then report their local sensing

© The Author(s) 2016
Y. Liao et al., *Listen and Talk*, SpringerBriefs in Electrical and Computer
Engineering, DOI 10.1007/978-3-319-33979-5_3

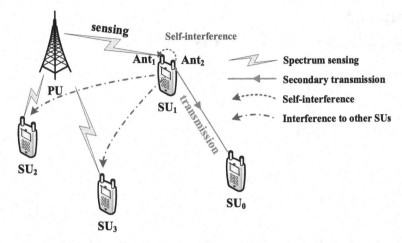

Fig. 3.1 System model of the cooperative sensing scenario, in which M SUs sense the same spectrum channel cooperatively, and report their local sensing results to a common fusion center (FC). The FC makes the final judgement and decides which SU should transmit in the next time slot

results to the FC. The FC then makes a final decision on the presence of the PU by combining the local reports and decides whether the PU is present in the current time slot.

3.1.1 Local Spectrum Sensing

For each SU_i in the network, the LAT protocol is adopted, i.e., Ant_{i1} is utilized to receive signal from the environment to sense the spectrum, and Ant_{i2} is utilized to transmit its own data. In each time slot, only one SU is allowed to transmit if the PU is absent. Without loss of generality, we assume only SU_1 is allowed to transmit, and all the other SUs keep silent.

For the secondary transmitter SU_1 itself, the spectrum sensing is almost the same as introduced in the previous sections. The received signal at Ant_{11} is written as

$$
\begin{aligned}
(\text{Silent } SU_1)\ y_1 &= \begin{cases} h_1 s_p + u_1, & \mathcal{H}_{01}, \\ u_1, & \mathcal{H}_{00}, \end{cases} \\
(\text{Active } SU_1)\ y_1 &= \begin{cases} h_1 s_p + w + u_1, & \mathcal{H}_{11}, \\ w + u_1, & \mathcal{H}_{10}, \end{cases}
\end{aligned}
\tag{3.1}
$$

where s_p denotes the PU's signal, $h_1 \sim \mathcal{CN}\left(0, \sigma_1^2\right)$ is the Rayleigh channel from the PU to Ant_{11}, $u_1 \sim \mathcal{CN}\left(0, \sigma_u^2\right)$ represents the complex-valued Gaussian noise, and $w \sim \mathcal{CN}\left(0, \chi^2 \sigma_s^2\right)$ is the RSI term, with σ_s^2 and χ^2 denote the power of SU$_1$'s transmit power and the suppression level of self-interference, respectively.

For any other SU$_i$ ($i \neq 1$), the signal from SU$_1$ is treated as interference. We have

$$\text{(Silent SU}_1)\ y_i = \begin{cases} h_i s_p + u_i, & \mathcal{H}_{01}, \\ u_i, & \mathcal{H}_{00}, \end{cases}$$

$$\text{(Active SU}_1)\ y_i = \begin{cases} h_i s_p + h_{1i} s_1 + u_i, & \mathcal{H}_{11}, \\ h_{1i} s_1 + u_i, & \mathcal{H}_{10}, \end{cases} \tag{3.2}$$

where s_1 is the SU$_1$'s signal, and $h_i \sim \mathcal{CN}\left(0, \sigma_i^2\right)$ and $h_{1i} \sim \mathcal{CN}\left(0, \sigma_{1i}^2\right)$ denote the channels from the PU and SU$_1$ to SU$_i$, respectively.

Again, energy detection is adopted as the sensing strategy, and the test statistics for local sensing at SU$_i$ can be expressed as

$$O_i = \frac{1}{N_s} \sum_{n=1}^{N_s} |y_i(n)|^2, \ i = 1, 2, \ldots, M. \tag{3.3}$$

Let $X = 0/1$ denotes silent/active state of SU$_1$. Given ϵ_{iX} as the detection thresholds at SU$_i$, the local probabilities of miss detection and false alarm are given by, respectively,

$$P_{im}^X(\epsilon_{iX}) = \Pr\left(O_i < \epsilon_{iX} | \mathcal{H}_{X1}\right), \tag{3.4}$$

$$P_{if}^X(\epsilon_{iX}) = \Pr\left(O_i > \epsilon_{iX} | \mathcal{H}_{X0}\right). \tag{3.5}$$

3.1.2 Data Report and Data Fusion

We assume the SUs report their one-bit hard decisions to the FC, and no error exists in the reporting process. In the FC, the OR fusion rule is adopted,[1] i.e., the FC decides the presence of the PU if at least one report declares that the PU is detected, and vise versa. The miss detection and false alarm probabilities of the cooperative decision is then given by

$$P_m^X = \prod_{i=1}^{M} P_{im}^X,$$

$$P_f^X = 1 - \prod_{i=1}^{M} \left(1 - P_{if}^X\right). \tag{3.6}$$

---

[1]Note that the FC can also use other fusion rules like the AND-rule or majority-rule. However, the main conclusions remain the same regardless of the specific choice of fusion rules. In this book, we adopt OR-rule as an representative.

3.1.3   Analysis of CSS in the LAT CRN

3.1.3.1   Local Sensing Error

Similar to the analysis in Sect. 2.2.2, the PDF of M_i can be approximated by a Gaussian distribution according to the CLT, and the PDF of the test statistics at any SU_i (O_i) can be derived based on (3.1) and (3.2). For simplicity, we assume that all channels are independent, and both the PU's and SU_1's signal is PSK modulated, and the power of the PU's signal is σ_p^2. Then the interference term $h_{1i}s_1$ ($i \neq 1$) can be treated as random complex Gaussian distributed, and the PDF of O_i ($i \neq 1$) and O_1 can be written in similar forms. The description and statistical properties are given in Table 3.1, where $\gamma_i = \frac{\sigma_i^2 \sigma_p^2}{\sigma_u^2}$, $\gamma_{1i} = \frac{\sigma_{1i}^2 \sigma_s^2}{\sigma_u^2}$ denote the SNR from the PU to SU_i and the interference-to-noise ratio (INR) caused by SU_1, respectively. Further, let $\gamma_{11} = \frac{\chi^2 \sigma_s^2}{\sigma_u^2}$ be the INR at SU_1, and the distribution of O_1 can be written uniformly with O_i ($i \neq 1$) as shown in Table 3.1. Due to space limitation, the mathematical derivation for Table 3.1 is omitted in this paper.

With the properties in Table 3.1, the local probabilities of miss detection and false alarm at SU_i can be derived from (3.4) and (3.5) as, respectively,

$$P_{im}^0 (\epsilon_{i0}) = 1 - Q\left(\left(\frac{\epsilon_{i0}}{(1 + \gamma_i)\sigma_u^2} - 1\right)\sqrt{N_s}\right),$$

$$P_{if}^0 (\epsilon_{i0}) = Q\left(\left(\frac{\epsilon_{i0}}{\sigma_u^2} - 1\right)\sqrt{N_s}\right),$$

$$P_{im}^1 (\epsilon_{i1}) = 1 - Q\left(\left(\frac{\epsilon_{i1}}{(1 + \gamma_i + \gamma_{1i})\sigma_u^2} - 1\right)\sqrt{N_s}\right),$$

$$P_{if}^1 (\epsilon_{i1}) = Q\left(\left(\frac{\epsilon_{i1}}{(1 + \gamma_{1i})\sigma_u^2} - 1\right)\sqrt{N_s}\right).$$

(3.7)

Table 3.1 Statistical properties of M_i

Hypothesis	PU	SU$_1$	$\mathbb{E}[M_i]$	var$[M_i]$
$\mathcal{H}_{00}$	Idle	Silent	σ_u^2	$\frac{\sigma_u^4}{N_s}$
$\mathcal{H}_{01}$	Busy	Silent	$(1 + \gamma_i)\sigma_u^2$	$\frac{(1+\gamma_i)^2 \sigma_u^4}{N_s}$
$\mathcal{H}_{10}$	Idle	Active	$(1 + \gamma_{1i})\sigma_u^2$	$\frac{(1+\gamma_{1i})^2 \sigma_u^4}{N_s}$
$\mathcal{H}_{11}$	Busy	Active	$(1 + \gamma_i + \gamma_{1i})\sigma_u^2$	$\frac{(1+\gamma_i+\gamma_{1i})^2 \sigma_u^4}{N_s}$

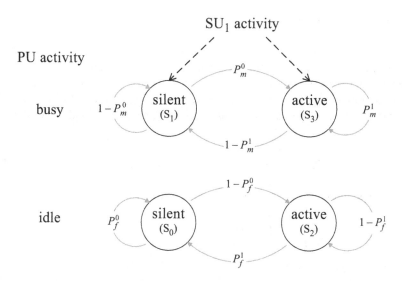

Fig. 3.2 Simplified state transition of the system

3.1.3.2   Collision Ratio and Spectrum Waste Ratio

Similar to the derivation in Sect. 2.2.2, we need to consider the state transition among the four states of the system. For simplicity, we assume that the probabilities that the PU changes its state within a slot μ, ν are sufficiently small, and we omit it in the consideration of state transition. Thus, the transition graph can be divided into two separated DTMCs as shown in Fig. 3.2.

Since we do not consider the state change of the PU within a slot, the collision ratio can be simplified as the conditional probability that the system stays at S_3 given that the PU is occupying the spectrum, i.e.,

$$P_c = \frac{P_3}{P_1 + P_3}. \tag{3.8}$$

Similarly, the spectrum waste ratio is given as

$$P_w = \frac{P_0}{P_0 + P_2}. \tag{3.9}$$

Considering the steady state probability, the collision ratio and spectrum waste ratio can be derived as, respectively,

$$P_c = \frac{P_m^0}{1 + P_m^0 - P_m^1};$$
$$P_w = \frac{P_f^1}{1 - P_f^0 + P_f^1}. \tag{3.10}$$

3.1.3.3 Local Sensing Thresholds

Substituting (3.7) to (3.6), and then to (3.10), we can obtain the collision ratio and spectrum waste ratio. With the collision ratio as the system constraint on the secondary network, the constraints of P_m^0 and P_m^1 can be given by (3.10), and then the constraints of P_{im}^X are given by (3.6). For simplicity, we set all miss detection probabilities P_{im}^X to be the same without further optimization. So with a fixed P_c, we have

$$P_{im}^X = (P_c)^{1/M}, \ \forall i = 1, 2, \ldots, M, \ X = 0, 1. \tag{3.11}$$

The corresponding thresholds ϵ_{iX} can be derived from (3.7) as

$$\epsilon_{i0} = \left(\frac{Q^{-1}\left(1 - (P_c)^{1/M}\right)}{\sqrt{N_s}} + 1 \right)(1 + \gamma_i)\,\sigma_u^2,$$

$$\epsilon_{i1} = \left(\frac{Q^{-1}\left(1 - (P_c)^{1/M}\right)}{\sqrt{N_s}} + 1 \right)(1 + \gamma_i + \gamma_{1i})\,\sigma_u^2.$$

We can see a rise in the thresholds when interference is introduced to the system. The local probabilities of false alarm can be derived accordingly:

$$P_{if}^0 (P_c) = Q\left(Q^{-1}\left(1 - (P_c)^{1/M}\right)(1 + \gamma_i) + \gamma_i\sqrt{N_s}\right),$$

$$P_{if}^1 (P_c) = Q\left(Q^{-1}\left(1 - (P_c)^{1/M}\right)\left(1 + \frac{\gamma_i}{1 + \gamma_{1i}}\right) + \frac{\gamma_i\sqrt{N_s}}{1 + \gamma_{1i}}\right). \tag{3.12}$$

Substituting (3.12) to (3.6) and (3.10), the spectrum waste ratio can be obtained.

Now consider a comparison between the performance of CSS in the LAT protocol and the non-cooperative sensing performance which could be obtained from (3.7) and (3.10). With cooperation, much larger local miss detection probabilities P_{im}^X are allowed at each cooperating SU as is shown in (3.11), and the corresponding local false alarm probabilities can be reduced sharply. Suppose we have ten cooperative SUs and the system P_m is set to be 0.01. Use (3.11) and we have $P_{im}^X > 63\%$ which is quite large and P_{if}^X could be depressed to a very low level. Also, the interference between two SUs can be much smaller than self-interference with careful selection of cooperative SUs, so the assistant sensing results can be more reliable than the result at the transmitting SU, and performance of CSS is quite likely to be much better than non-cooperative sensing in the LAT CRNs.

3.1.3.4 Secondary Throughput

Once a spectrum hole is detected, SU_1 begins transmitting data to SU_0. Only Ant_{12} is used for transmission and the secondary throughput can be measured as

$$
\begin{aligned}
C &= (1 - P_w) \log_2 \left(1 + \frac{\sigma_s^2 \sigma_t^2}{\sigma_u^2} \right) = (1 - P_w) \log_2 (1 + \gamma_t) \\
&= \frac{1 - P_f^0}{1 - P_f^0 + P_f^1} \cdot \log_2 (1 + \gamma_t),
\end{aligned}
\tag{3.13}
$$

where σ_t^2 is the variance of the Rayleigh channel from SU_1 to the receiver SU_0, $\gamma_t := \frac{\sigma_s^2 \sigma_t^2}{\sigma_u^2}$ is the SNR in transmission, and P_f^0 and P_f^1 are given earlier.

Note that the two factors in (3.13) are both related to transmit power σ_s^2. When σ_s^2 increases, on the one hand, the INRs (γ_i) increase and P_f^1 rises accordingly, while on the other hand, the achievable sum rate $\log_2 (1 + \gamma_t)$ increases. Thus, the tradeoff between transmit power and throughput may also exist as analyzed in Chap. 2.

3.1.4 Comparison with Other Protocols

In this section, we compare the proposed cooperative LAT protocol with cooperative/non-cooperative LBT protocols and non-cooperative LAT protocol. We first provide a sketch on how to derive the performance of the cooperative LBT protocol, and show the comparison by simulation results.

3.1.4.1 Cooperative Spectrum Sensing in LBT Protocol

In cooperative LBT protocol, local spectrum sensing of each SU can be regarded completely independent. Thus, the local spectrum sensing is different from the non-cooperative protocol only in the constraint of local miss detection probability P_{im}'. Since the performance of non-cooperative LBT protocol has been analyzed in Sect. 2.2.3.3, we only provide a brief sketch and main conclusions of the cooperative LBT protocol here.

Proposition 3.1. *In cooperative LBT protocol, the spectrum waste ratio is given by*

$$
P_w' = 1 - (1 - \lambda) \prod_{i=1}^{M} \left(1 - Q \left(Q^{-1} \left(1 - \left(\frac{P_c}{1 - \lambda} \right)^{1/M} \right) \eta_i + \gamma_i \sqrt{2\lambda N_s} \right) \right),
\tag{3.14}
$$

where $\lambda = T_s/T$ *denotes the ratio of sensing duration over the whole slot length, and* $\eta_i := \left((\beta_i\gamma_i)^2 + (\gamma_i + 1)^2\right)^{1/2}$, *with* β_i *denoting the spatial correlation coefficient of the two antennas on* SU_i.

Proof. The sensing performance of the conventional cooperative LBT protocol can be derived through similar procedure in Sect. 2.2.3.3. The constraint of local miss detection probability is

$$P'_{im} = \left(\frac{P_c}{1-\lambda}\right)^{1/M}, \tag{3.15}$$

where $\lambda = T_s/T$ denotes the ratio of sensing duration over the whole slot length. The local false alarm probability can be derived via (2.53) as

$$P'_{if}(P_c;\lambda) = \mathcal{Q}\left(\mathcal{Q}^{-1}\left(1 - \left(\frac{P_c}{1-\lambda}\right)^{1/M}\right)\eta_i + \gamma_i\sqrt{2\lambda N_s}\right), \tag{3.16}$$

The spectrum waste ratio can then be given as

$$\begin{aligned}
P'_w &= \lambda + (1-\lambda)\left(1 - \prod_{i=1}^{M}\left(1 - P'_{if}\right)\right) \\
&= 1 - (1-\lambda)\prod_{i=1}^{M}\left(1 - \mathcal{Q}\left(\mathcal{Q}^{-1}\left(1 - \left(\frac{P_c}{1-\lambda}\right)^{1/M}\right)\eta_i + \gamma_i\sqrt{2\lambda N_s}\right)\right).
\end{aligned} \tag{3.17}$$

$\square$

The secondary throughput of the cooperative LBT protocol can be obtained from (2.56) by replacing P'_w given in (2.44) by that given in (3.14).

3.1.4.2 Results

Now, we evaluate the performance of cooperative/non-cooperative LAT and LBT protocols with simulation results. Simulation parameters are listed in Table 3.2. For simplicity, in the simulation of the conventional cooperative/ non-cooperative LBT protocols, we set the spatial correlation of each SU to be the same, i.e., $\beta_i = \beta, \forall i = 0, 1, 2, \ldots, M$. The SNRs from PU to SUs are set to be uniform distributed as $\mathcal{U}\left([0.95\bar{\gamma}, 1.05\bar{\gamma}]\right)$, and the SNR in Fig. 3.3 means the mean value $\bar{\gamma}$.

In Fig. 3.3, we study the performance of the cooperative/non-cooperative LAT/LBT protocols under different average sensing SNR. Lines in this figure are based on analytical expressions (3.13) and (2.56), and the asterisks are the numerical results, which match perfectly. It shows that both the cooperative LAT and cooperative LBT perform better than non-cooperative ones when average sensing SNR is small, and the maximum achievable throughput, i.e., the throughput under

Table 3.2 Simulation parameters

Parameters	Value
The number of samples in each slot (N_s)	300
Ratio of sensing time in the LBT protocol (λ)	1/4
Number of cooperative SUs (M)	10
The interference channels from SU$_1$ to others ($|\sigma_{1i}|$)	$\mathcal{U}([0, 0.25])$
The RSI factor (χ^2)	0.01
The spatial correlation coefficient (β)	0.8
Normalized SU$_1$'s transmit power (σ_s^2/σ_u^2)	10 dB
Collision ratio (P_c)	0.01

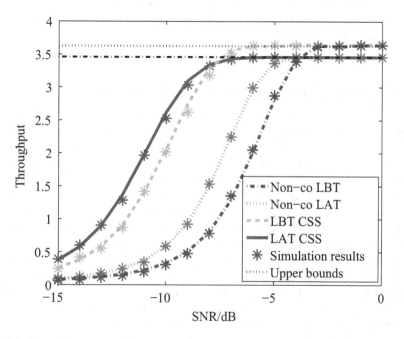

Fig. 3.3 Secondary throughput versus sensing SNR γ_s, in which the probability of the PU's arrival $\mu = 1/500$, departure $\nu = 6/500$, the collision ratio is 0.1, the sample number of a slot N_s is 300, normalized secondary transmit power $\sigma_s^2/\sigma_u^2 = 10$dB, the RSI factor χ^2 in the LAT varies between 0.1 and 0.01, the spatial correlation coefficient β is 0.85, and the ratio of sensing duration changes between 1/3 and 1/10

perfect sensing of both protocols does not change for cooperation. Under low sensing SNR, the proposed LAT protocols, both non-cooperative and cooperative, tend to perform higher throughput due to longer sensing time and lower false alarm probability and approach the maxima earlier. When sensing SNR is high enough and the sensing performances become quite reliable under all protocols, the throughput is largely decided by the maximum achievable throughput as is shown by the dotted black lines.

3.2   Extension 2: Dynamic Spectrum Access

In this section we consider the scenarios where multiple FD SUs contend for the same licensed band. In these scenarios, how to design proper dynamic spectrum access (DSA) strategies to fully explore the advantage of FD techniques becomes an interesting problem. There are mainly two kinds of DSA strategies, namely distributed and centralized strategies, which we will discuss respectively in the remainder of this section. In distributed DSA, each SU needs to sense and decide whether to access the spectrum independently without a central controller. For this scenario, a design of a new access and backoff mechanism is presented in Sect. 3.2.1. In centralized DSA, there exists a central control that allocates time and spectrum resources to SUs. For this case, we discuss the power and spectrum allocation of a FD cognitive AP system in Sect. 3.2.2.

3.2.1   Distributed Dynamic Spectrum Access

In many scenarios such as in ad hoc CRNs where the SUs compete for several PU channels, deploying a central controller is not always possible. Therefore, distributed DSA will be required, by which each SU has to independently gather, exchange, and process the information of the wireless environment. The commonly used CSMA/CA in distributed DSA with HD users can effectively reduce collision probability [4], but some problems still exist: (1) collision among the SUs can never be detected if the SUs are synchronized, such that the secondary transmission may fail to a large scale, and (2) SUs cannot abort transmission when collision happens, which leads to long collision duration. With FD-CR, the SUs can not only detect the presence of PUs, but also detect collision with other SUs during transmission, such that the collision duration is reduced significantly. But the RSI may degrade the collision detection accuracy, which cannot be ignored. In the rest of this section, we present a feasible distributed DSA protocol for FD-CRNs, and show its effectiveness of achieving high spectrum utilization ratio when the number of users goes up.

3.2.1.1   System Model

The system considered in this section is same as that in Sect. 3.1, i.e., a CRN consisting of one PU with a licensed channel and M FD-enabled SUs, where each SU can sense the channel and transmit simultaneously, and they are allowed to access the spectrum only when the PU is absent. The difference is that the SUs do not cooperate with each other, instead, they perform spectrum sensing individually and access the spectrum according to a certain mechanism without communication with other SUs.

The PU's usage of the channel is modeled as a non-slotted alternating "ON/OFF" random process, and both the arrival and departure processes of the PU are assumed to be Poisson. In this case, the distributions of busy and idle time are exponential. We denote the average lengths of the idle periods and occupied periods of the PU as t_0 and t_1, respectively, and the minimum time for a SU to detect the occupancy of the channel as T. Considering the common case that the PU's state changes sufficiently slowly compared to SU's sensing time T, we assume that $t_0, t_1 \gg T$. Let $l_0 = \frac{t_0}{T}$ and $l_1 = \frac{t_1}{T}$, and $a = \frac{l_0}{l_0 + l_1}$ is the percentage that the spectrum is free of the PU signal.

Each SU adopt LAT protocol for local spectrum sensing, which has been proposed and analyzed in the previous sections. The traffic of SUs is slotted with duration T. At the end of each slot, each SU combines the detected signal in the slot and makes a decision about whether to transmit or not in the next slot. Since the RSI exists in the sensing of the transmitting SUs, the sensing of these SUs cannot be assumed perfect, which is different from the common perfect-sensing assumption in conventional half-duplex scenarios. For simplicity, we omit the noise term in (2.5) and (2.4). Thus, the sensing of a silent SU becomes perfect, while for a transmitting SU, the sensing is imperfect with false alarm probability P_f and miss detection probability P_m. Furthermore, consider that if several (more than two) SUs collide in a slot, the received signal in sensing would be much larger than the RSI, we assume that the miss detection probability only applies to the case when two SUs collide or one SU collides with the PU.

3.2.1.2 FD-DSA Protocol

In this part, we present a feasible distributed DSA mechanism for FD SUs that can effectively eliminate collision between SUs and the PU as well as collision among SUs.

Sensing with FD With FD techniques, SUs can keep sensing during transmission. The traffic of SUs is slotted with duration T, given that the SUs cannot make any decision with duration shorter than T. As shown in Fig. 3.4, at the end of each slot, each SU combines the detected signal of the slot and make a decision about whether to transmit or not in the next slot. On one hand, if SUs detect the PU's signal or other SUs' signal, they keep silent and only sense the spectrum in the next slot. On the other hand, if the spectrum is sensed idle, SUs do not access the spectrum immediately. Instead, each of them randomly chooses one duration from the set of *contention free periods (CFP)*, and waits for the chosen duration before they transmit. Note that all SUs keep sensing when they wait to transmit, and once the PU's signal or other SU's signal is detected, the waiting period suspends. The waiting period resumes when the spectrum is sensed idle again.

Contention Window For simplicity, the CFP is also slotted with duration T, and the contention window length is randomly chosen from $C = \{0, T, 2T, 3T, \ldots, (W-1)T\}$, i.e., the size of the CFP is W. At the beginning of a slot, if the spectrum is

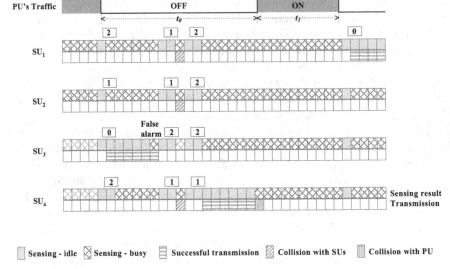

Fig. 3.4 DSA protocol for FD-CRNs, where there are four SUs, and the contention window length is 3

sensed idle in the previous slot, and the countdown of the waiting time has reached zero, the transmission begins. The transmission of any SU will not cease until other users are detected utilizing the band. Then, if another spectrum hole is detected, the SUs that have just performed transmission reselect a contention period randomly and start countdown, while other SUs continue the uncompleted countdown without another selection. This mechanism guarantees that every SU can count to zero within $(W - 1)$ slots and begins transmission.[2]

3.2.1.3 Performance Analysis

In this part, we analyse the performance of the proposed DSA protocol for FD-CRNs. First we check the collision percentage between primary and secondary networks. Then we calculate the spectrum utilization ratio, which is one of the most important metrics to evaluate the protocol performance. Also, we propose an algorithm to obtain the optimal contention window length.

Modeling the Imperfect Sensing Caused by the RSI As mentioned in the previous sections, the RSI degrades sensing performance when a SU is transmitting. For simplicity, in the analysis in this part, we omit the noise term in the received

---

[2]Note that the contention scheme does not need to be the scheme above. Other schemes such as that with changeable contention windows can also be adopted. The expressions of performance may be different but the major conclusions remain the same.

signal for sensing, and the sensing when SU is silent becomes perfect. For a transmitting SU, the sensing is imperfect. We assume that if a SU is transmitting, the probability of false alarm, i.e., it judges that the band is utilized by other users when it is actually not, is P_f, and the miss detection probability, i.e., the probability that it fails to detect others' signal, is P_m. Furthermore, when several (more than two) SUs collide in the same slot, the received signal in sensing would be much larger than the RSI, and we assume that the collision can be detected with probability 1, i.e., the miss detection probability only applies to the case when two SUs collide or one SU collides with the PU.

Collision with the Primary Networks As we assumed before, if a SU is silent, it can sense the PU's signal with probability 1. Also, if more than one SUs are transmitting and PU is busy, all SUs can detect the interference and stop transmitting in the next slot. Thus, the case that SUs collide with PUs only happens at the beginning of PU's arrival when some SUs are transmitting. And only when one SU collides with the PU, the collision length may be longer than one slot. It is shown from the analysis above that the average collision length with the PU is slightly longer than one secondary slot. With the assumption that the PU changes sufficiently slowly compared with the secondary slot length, this collision length is quite negligible.

Spectrum Usage We use the spectrum utilization ratio ρ to evaluate the performance of the protocol. The utilization ratio is defined as

$$\rho = \frac{\mathbb{E}[k]}{\mathbb{E}[k] + \mathbb{E}[\kappa]}, \qquad (3.18)$$

where $\mathbb{E}[k]$ is the average continuous transmit length, and $\mathbb{E}[\kappa]$ is the average spectrum waste length before a successful transmission. Since the average continuous transmit length and waste length is independent, we calculate them separately in the rest of this part.

1. **Duration of the continuous transmission:** If a SU begins transmission without collision with other SUs or the PU network, there are only two causes that stop the transmission: (1) the PU's arrival, and (2) a false alarm in its own sensing. The probability that the PU comes in a slot is $1 - e^{-1/l_0}$, and the false alarm probability is P_f. Thus, the probability that the SU continues its transmission in the next slot is

$$P_t = e^{-1/l_0}(1 - P_f) + (1 - e^{-1/l_0})P_f, \qquad (3.19)$$

and the average continuous transmit length can be calculated as

$$\mathbb{E}[k] = \sum_{k=1}^{\infty} P_t^{k-1} k = \frac{1}{(1 - P_t)^2}. \qquad (3.20)$$

2. **Spectrum waste:** The spectrum waste mainly consists of three parts: (i) pure sensing—after the transmission of the PU or SUs, all the SUs need to sense for one slot to detect the idleness of the spectrum; (ii) contending—when the spectrum is sensed idle, and all SUs are waiting to transmit; (iii) collision between SUs leads to inefficient transmission. First we consider case (i) and (ii) where there is no collision. Then we deal with the collision case (iii).

- If there is no collision before one SU begins successful transmission, it means that the CFP of this SU is the single minimum among all SUs. The probability that the waiting time is κ_1 is given by

$$P_1(\kappa_1) = \frac{M}{W}\left(1 - \frac{\kappa_1}{W}\right)^{M-1}, \quad \kappa_1 = 1, 2, \ldots, W - 1. \tag{3.21}$$

Here we omit the slot length T in the derivation for simplicity. The average waste length in this case can be calculated as

$$
\begin{aligned}
\mathbb{E}\left[\kappa_1\right] &= \sum_{\kappa_1=1}^{W-1} P_1(\kappa_1) \cdot \kappa_1 = \sum_{\kappa_1=1}^{W-1} \frac{M}{W}\left(1 - \frac{\kappa_1}{W}\right)^{M-1} \kappa_1 \\
&= \sum_{i=1}^{W-1} \frac{M}{W}\left(\frac{i}{W}\right)^{M-1}(W - i).
\end{aligned}
\tag{3.22}
$$

- If the collision happens before a successful transmission, there are three circumstances: (i) more than two SUs collide and they will stop transmission in the next slot; (ii) two SUs collide and they stop together during some slots; (iii) two SUs collide, one of them stops first, and the other perform successful transmission after that. For case (i), the probability that the waiting time before the first collision equals κ_{21} is

$$
\begin{aligned}
P_{21}(\kappa_{21}) &= \left(1 - \frac{\kappa_{21} - 1}{W}\right)^M - \frac{M}{W}\left(1 - \frac{\kappa_{21}}{W}\right)^{M-1} \\
&- \frac{M(M-1)}{2W^2}\left(1 - \frac{\kappa_{21}}{W}\right)^{M-2} - \left(1 - \frac{\kappa_{21}}{W}\right)^M,
\end{aligned}
\tag{3.23}
$$

and the average waste length is

$$\mathbb{E}\left[\kappa_{21}\right] = \sum_{\kappa_1=1}^{W} P_{21}(\kappa_{21})(\kappa_{21} + \mathbb{E}[\kappa]). \tag{3.24}$$

If two SUs begin transmission together, the occurrence probability of case (ii) is

$$P_2 = \sum_{i=0}^{\infty} P_m^{2i}(1 - P_m)^2 = \frac{1 - P_m}{1 + P_m}, \tag{3.25}$$

and the occurrence probability of case (iii) is $P_3 = \frac{2P_m}{1+P_m}$. The probability of waiting for κ_2 is

$$P_{22}(\kappa_2) = \frac{M(M-1)}{2W^2}\left(1 - \frac{\kappa_2}{W}\right)^{M-2}, \quad \kappa_2 = 1, 2, \ldots, W-1. \tag{3.26}$$

The average waste length for case (ii) can be expressed similarly as in case (i):

$$\mathbb{E}[\kappa_{22}] = \sum_{\kappa_{22}=1}^{W-1} P_2 P_{22}(\kappa_{22})\left(\tilde{k} + \kappa_{22} + \mathbb{E}[\kappa]\right), \tag{3.27}$$

where $\tilde{k}$ is the average collide length when two SUs collide, which can be calculated as $\frac{1}{1-P_m^2}$.

For case (iii), the expectation of the waste length is

$$\mathbb{E}[\kappa_{23}] = \sum_{\kappa_{23}=1}^{W-1} P_3 P_{22}(\kappa_{23})\left(\tilde{k} + \kappa_{23}\right). \tag{3.28}$$

Thus, the expectation of the overall waste length can be calculated as

$$\mathbb{E}[\kappa] = \mathbb{E}[\kappa_1] + \mathbb{E}[\kappa_{21}] + \mathbb{E}[\kappa_{22}] + \mathbb{E}[\kappa_{23}]. \tag{3.29}$$

Substituting (3.22), (3.24), (3.27) and (3.28) into (3.29), we can calculate the waste length expectation as

$$\mathbb{E}[\kappa] = \frac{1 + S(M) + \frac{M(M-1)}{2W^2}\left(\tilde{k} - \frac{2P_m}{1+P_m}\right)S(M-2)}{\frac{M}{W}S(M-1) + \frac{M(M-1)}{2W^2}\frac{2P_m}{1+P_m}S(M-2)}, \tag{3.30}$$

where $S(m) = \sum_{i=1}^{W-1}\left(\frac{i}{W}\right)^m$.

The channel usage ratio can be readily derived by substituting (3.20) and (3.30) into (3.18).

Optimal Contention Window   Now that only $\mathbb{E}[\kappa]$ in (3.18) is relevant to the contention window length, we focus on the average waste length to discuss the impact of the contention window length on the spectrum usage ratio.

If the contention window is too small, the probability of collision among SUs is large, and there may be long recurrence of "wait–collide–sensing–wait", and the spectrum usage efficiency is low. On the other hand, if the contention window is too large, the collision ratio is small; yet the average contending time (before the smallest CFP reaches zero) increases proportionally to the maximum window length, and the usage efficiency also gets degraded. Therefore, there exists an optimal contention window size that minimizes the average waste time before a successful transmission.

Since the form of (3.30) is complicated, and the optimal point is difficult to calculate, we approximate the summation $S(m)$ by the following integral:

$$S(m) \approx \int_0^{1-\frac{1}{W}} x^m dx = \frac{1}{m+1}\left(1 - \frac{1}{W}\right)^{m+1} \approx \frac{1}{m+1} - \frac{1}{W}.$$

The original summation approaches the above expression as the contention window size increases.

Thus, (3.30) can be simplified as

$$\mathbb{E}[\kappa] \approx \frac{A - 1/W + B/W^2 - C/W^3}{D/W - E/W^2 - F/W^3}, \qquad (3.31)$$

where A, B, C, D, E, and F are positive parameters, which can be derived from (3.30). By solving $\frac{d\mathbb{E}[\kappa]}{dW} = 0$, the optimal contention window length can be obtained.

3.2.1.4 Comparison with DSA in HD-CRNs

In conventional DSA in HD-CRNs, an SU performs sensing before transmission, and the sensing is commonly assumed perfect[4]. In the conventional system, collision typically happens in the following two cases: (1) the PU comes while SUs are transmitting, and this collision cannot be detected until the next sensing period; and (2) when SUs are synchronized, they cannot detect collision among them since all active SUs stop transmission and sense the channel at the same time, and the collision will last during the whole transmission. Consider the maximum utilization of the conventional DSA protocol, which is under the constraint of the sensing and transmission duration, can be written as

$$\rho_{HD} < \frac{\text{transmitting time}}{\text{sensing time} + \text{transmitting time}}. \qquad (3.32)$$

Thus, if we set the average collision duration of the proposed FD-CRN based DSA and the conventional HD-CRN based DSA as the same, i.e., the collision duration is about one slot, the transmitting period cannot be longer than two slots, which means that the utilization ratio of the conventional protocol cannot reach 70%. However, in the proposed protocol, the utilization ratio does not have this kind of hard constraint. We will show in the simulation section that the ratio can approach 90 % with the proposed protocol.

Also, in the conventional protocol, if all SUs are synchronized and they have the slot structure, when several SUs collide with each other, the collision will never be detected by themselves, and the collision may last for quite long time and severely deteriorate the performance. While in the DSA for FD-CRNs, whenever several SUs collide, they will detect the collision in one slot with a large probability and stop transmission in the next to avoid longer ineffective transmission. In this aspect, the proposed protocol largely outperforms the conventional one.

3.2.1.5 Results

In Fig. 3.5, we set the PU's arrival rate as 0.001, the false alarm probability as 0.4 to shorten the average continuous transmit length, and the total simulation length is 1,000,000 slots. We show in Fig. 3.5 the results of the average waiting length, which have been calculated in Sect. 3.2.1.3. The analytical results (the solid and dotted lines) are based on the simplified expression of $\mathbb{E}[\kappa]$ in (3.31), which matches the simulation results when the contention window length W is sufficiently long. Also, we can find that the optimal contention window length calculated from the simplified $\mathbb{E}[\kappa]$ represents the real case well, shown by the vertical lines. Comparing the solid and dashed lines, when other parameters remain the same, the

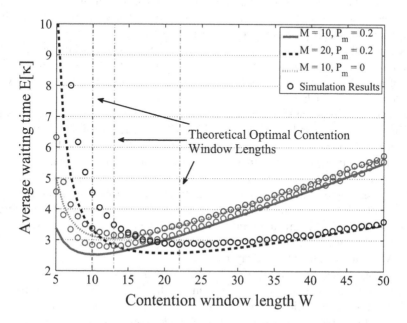

Fig. 3.5 DSA in FD-CRNs: average waiting length vs. contention window length

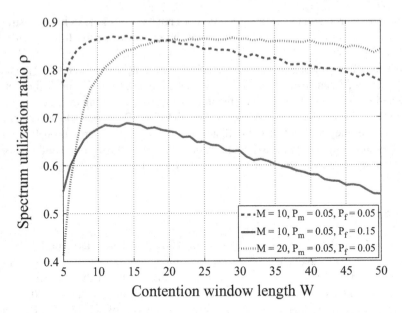

Fig. 3.6 DSA in FD-CRNs: spectrum utilization ratio vs. contention window length

scenario with more SUs is more adapted to larger contention windows, while short contention windows suits the case with fewer SUs. Another interesting point is the comparison between the solid and dotted lines. When miss detection probability rises, the average wait time decreases. The reason is that when two SUs collide with each other, the probability that one SU stops before the other, i.e., $\frac{2P_m}{1+P_m}$, increases with the miss detection probability, and this leads to shorter waste time before successful transmission. However, even though not shown in the figure, the rise of miss detection probability inevitably results in a longer collision time with the PU network, which is harmful to the whole system.

Figure 3.6 shows the spectrum utilization ratio of the proposed protocol under different contention window length, in which the PU's arrival rate is fixed on 0.0001, and miss detection probability is 0.05. Figure 3.6, as well as Fig. 3.5 shows that the optimal contention window length increases with the number of SUs, and it typically several slots longer than the number of SUs. Also, it is shown in Fig. 3.6 that the spectrum utilization ratio is affected by the false alarm probability more than the SU's number or contention window length, i.e., when false alarm probability rises from 0.05 to 0.15, the utilization ratio drops from the dashed line to the solid line significantly. This indicates the importance of improving sensing performance of the LAT protocol in the physical layer.

When we check the performance when the number of SUs changes, we find that the maximum achievable spectrum utilization ratio is almost the same when the contention window varies among several values, as is shown in Fig. 3.7. And the utilization ratio can reach around 87 % under careful matching of the SU's

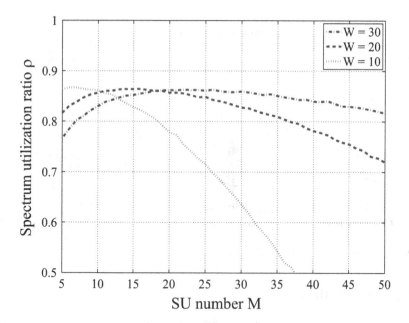

Fig. 3.7 DSA in FD-CRNs: spectrum utilization ratio vs. number of secondary users

number and contention window length. From the other perspective, with the increase of the number of SUs, the total spectrum utilization ratio can remain at a high level with carefully designed length of the contention window under the proposed decentralized DSA protocol for FD-CRNs.

3.2.2 Centralized Dynamic Spectrum Access

In this section, we consider the scenario where there exists a central controller to schedule and monitor the access and transmission of all SUs. Specifically, we consider a cognitive cellular network (CCN), in which a secondary base station (SBS) is deployed to control the secondary transmissions between itself and multiple SUs. The SBS, as a central controller, is in charge of sensing the primary spectrum and deciding frequency bands and power of each secondary transmission. The SUs are fully controlled by the SBS and no direct transmissions are allowed between each other. Here, we only consider the downlink transmissions from the SBS to the SUs. In traditional CCNs using the LBT protocol, the spectrum sensing and spectrum access exclusively compete for the same radio resource. Thus, most studies have focused on scheduling the time and order of sensing and transmission, so as to achieve the optimal tradeoff [5, 6, 14]. However, with the help of FD technique, i.e., when the SBS has strong self-interference suppression (SIS) capability, the whole scenario can be significantly changed.

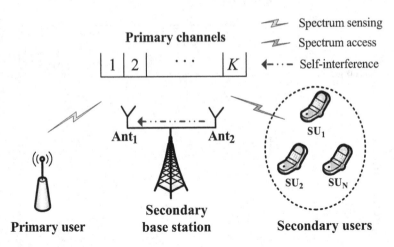

Fig. 3.8 System model of a full-duplex cognitive cellular network

3.2.2.1   System Model

Consider a network with one PU, one SBS and M SUs, as shown in Fig. 3.8. The primary network is an OFDM system, in which the PU transmits on K orthogonal channels. On each channel, the probabilities that the PU is present and absent are given by P_{on} and $P_{off} = 1 - P_{on}$, respectively. The secondary network is a spectrum overlay-based cognitive cellular network consisting of one SBS and M SUs. The SBS, equipped with two antennas, is a full-duplex device with strong SIS capability. When the SBS transmits to the SUs using Ant_2, the self-interference can be deeply suppressed such that the SBS can simultaneously sense the primary signal using Ant_1. Once the idle channels are detected, the SBS decides the power allocation and spectrum access strategy for the downlink transmissions of the M SUs. At antenna Ant_1, the received signal on channel k is given by:

$$y_k = \begin{cases} h_k s_{p,k} + w_k + u_k, & \text{ON}, \\ w_k + n_k, & \text{OFF}, \end{cases} \tag{3.33}$$

where $h_k s_{p,k}$ is the received signal of the PU, w_k is the self-interference leaked from antenna Ant_2, and u_k is the noise signal. We denote by $\sigma_u^2 = \mathbb{E}\{u_k^2\}$ as the noise power and $\gamma_{s,k} = \mathbb{E}\left[h_k^2 s_{p,k}^2\right]/\mathbb{E}\left[n_k^2\right]$ as the received SNR of the PU. The SIS capability of the SBS is quantified by χ^2, which represents the ratio between the RSI and the transmitting power. On channel k with transmitting power P_k, the self-interference-to-noise-ratio (INR) is denoted by $\gamma_{i,k} = \mathbb{E}\left[w_k^2\right]/\mathbb{E}\{u_k^2\} = \chi^2 P_k/\sigma_u^2$.

Using an energy detector, the miss detection and false alarm probabilities of the SBS on channel k are given by:

$$P_{m,k}(P_k, \epsilon_k) = 1 - Q\left(\left(\frac{\epsilon_k}{1 + \gamma_{s,k} + \gamma_{i,k}} - 1\right)\sqrt{N_s}\right), \tag{3.34}$$

and

$$P_{f,k}(P_k, \epsilon_k) = Q\left(\left(\frac{\epsilon_k}{1 + \gamma_{i,k}} - 1\right)\sqrt{N_s}\right), \tag{3.35}$$

where ϵ_k is the threshold on channel k and N_s is the number of samples.

Let β_k be a binary variable denoting the presence and absence of the PU on channel k, where $\beta_k = 0$ represents that the PU is present and $\beta_k = 1$ represents the absence. Let $\alpha = \{\alpha_{k,m}\}$ be a $K \times M$ binary matrix denoting the spectrum access strategy of the SBS, where $\alpha_{k,m} = 1$ represents that channel k is used to transmit to SU m and $\alpha_{k,m} = 0$ represents the opposite. We assume that each SU can occupy at most one channel and the total transmit power of the SBS is P. Thus, in order to maximize the total throughput of the SBS, the optimization problem is described as follows:

$$\max_{\{\alpha_{k,m}\},\{P_k\}} \sum_{k=1}^{K}\sum_{m=1}^{M} \beta_k\alpha_{k,m}(1 - P_{f,k})\log\left(1 + \frac{P_k|h_{k,m}|^2}{\sigma^2}\right), \tag{3.36a}$$

$$s.t.\ P_{m,k} \leq P_m,\ k = 1, 2, \ldots, K, \tag{3.36b}$$

$$\sum_{k=1}^{K} P_k \leq P, \tag{3.36c}$$

$$\sum_{k=1}^{K} \alpha_{k,m} \leq 1,\ m = 1, 2, \ldots, M, \tag{3.36d}$$

$$\sum_{m=1}^{M} \alpha_{k,m} \leq 1,\ k = 1, 2, \ldots, K, \tag{3.36e}$$

where $h_{k,m} \sim \mathcal{CN}(0, 1)$ denotes channel k for SU m, (3.36b) requests an upper bound P_m for the miss detection probability to ensure the PU's outage probability constraint, (3.36c) is the total power constraint of the SBS, (3.36d) ensures that each SU can occupy at most one channel, and (3.36e) ensures that each channel is exclusively occupied by at most one SU.

As we see in (3.36), the joint spectrum access and power allocation problem is to find an optimal combination of channels, users and power, so as to maximize the total throughput, and at the same time, satisfy all constraints. We will see that this problem can be seen as a three-dimensional matching problem in graph theory.

3.2.2.2 Joint Spectrum Access and Power Allocation

In this part, we reconsider problem (3.36) from the perspective of *three-dimensional matching* and prove its NP-hardness to find the optimal solution. Then, we show the influence of power allocation and present an approximate solution by extending a well-known *two-dimensional matching* algorithm.

NP-Hardness of the Optimal Solution Firstly, we show the equivalence of the problem (3.36) and the three-dimensional matching problem to prove the NP-hardness of the considered optimization problem.

Definition 3.1 (Three-Dimensional Matching). For a set $T \subseteq X \times Y \times Z$ of ordered triples where X, Y and Z are disjoint sets, each element $t \in T$ has an associated weight $w(t) \in \mathbb{R}$. A subset $S \subset T$ is a three-dimensional matching if for any two distinct elements $t_1 = (x_1, y_1, z_1), t_2 = (x_2, y_2, z_2) \in S$, we have $x_1 \neq x_2, y_1 \neq y_2$ and $z_1 \neq z_2$. We denote by Ψ as the set of all three-dimensional matchings, and the problem is to find the matching with the maximal weight, that is, $S^* = \arg\max_{S \in \Psi} \{\sum_{t \in S} w(t)\}$.

Proposition 3.2. *The optimization problem (3.36) is not easier than a three-dimensional matching problem.*

Proof. Let $X = \{1, 2, \ldots, K\}$ denote the K channels of the PU's spectrum, and let $Y = \{1, 2, \ldots, M\}$ denote the M SUs of the SBS. We assume that the optimal power allocation of channel k is given by P_k^* and let $Z = \{P_1^*, P_2^*, \ldots, P_K^*\}$ denote the set of all optimal powers. Consider a three-dimensional matching where $T = X \times Y \times Z$, and $w((k, m, P_l^*))$ is the maximal secondary throughput of SU m on channel k with transmit power P_l^*, given by

$$\max_{\epsilon_k : P_{m,k}(P_l^*, \epsilon_k) \leq P_m} \left\{ \beta_k \left[1 - P_{f,k}(P_l^*, \epsilon) \right] \log \left(1 + \frac{P_l^* |h_{k,m}|^2}{\sigma_u^2} \right) \right\}. \tag{3.37}$$

Thus, the optimal solution $\{\alpha_{k,m}^*\}$ and $\{P_k^*\}$ of (3.36) provides the optimal solution S^* of the three-dimensional matching, where $(k, m, P_l^*) \in S^*$ if and only if $\alpha_{k,m}^* = 1$ and $k = l$, as shown in Fig. 3.9. Therefore, the optimization problem (3.36) is no easier than a three-dimensional matching problem. $\square$

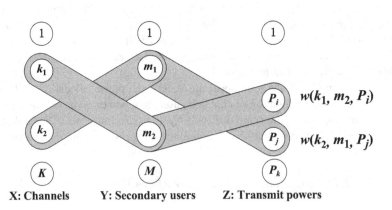

Fig. 3.9 The three-dimensional matching for the considered optimization problem

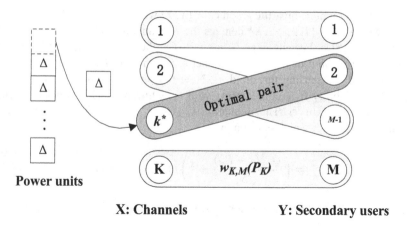

Fig. 3.10 The proposed algorithm based on two-dimensional matching

The three-dimensional matching problem is known to be NP-complete [7], which means that the joint spectrum access and power allocation problem (3.36) must be NP-hard, and it cannot be solved optimally in polynomial time.

Approximate Algorithm Although the three-dimensional matching is NP-complete, there are efficient polynomial-time algorithms for finding a two-dimensional matching, for example, the Kuhn-Munkres algorithm [8] and the Hopcroft-Karp algorithm [9]. We now propose an approximate algorithm based on the two-dimensional matching, as in Fig. 3.10. The idea is consider the spectrum access as a two-dimensional matching problem between the channels and the SUs. The total transmit power P is equally divided into M discrete units with each unit $\Delta = P/M$, and the discrete units are iteratively allocated by the SBS. In each iteration, one power unit is added to the channel that generates the largest marginal throughput of the current SU, and then, the two-dimensional matching method is performed to optimize the spectrum access strategy with the new power allocation. The algorithm stops when all power units are allocated or there is no channel-user pair that can increase its secondary throughput by increasing the transmit power.

Definition 3.2 (Two-Dimensional Matching). For a set $T \subseteq X \times Y$ of pairs where X and Y are disjoint sets, each pair $t = (x, y) \in T$ has an associated weight $w_{x,y} \in \mathbb{R}$. A subset $S \subset T$ is a two-dimensional matching if for any two distinct elements $t_1 = (x_1, y_1), t_2 = (x_2, y_2) \in S$, we have $x_1 \neq x_2$ and $y_1 \neq y_2$. We denote by Φ as the set of all two-dimensional matchings, and the problem is to find the maximum weighted matching, that is, $S^* = \arg\max_{S \in \Phi}\{\sum_{t \in S} w(t)\}$. Note that the two-dimensional matching problem can be seen as a graph theory problem. Given a bipartite graph $G = (X, Y)$ where $V(G) = X \cup Y$ and $E(G) = T$, and the weights $w_{x,y}$ of all edges $(x, y) \in E(G)$, a matching is defined as a set of edges that do not coordinate with each other, and the problem is to find the maximum weighted matching in graph G.

Consider a complete bipartite graph $G = (X, Y)$ with $V(G) = X \cup Y$ and $E(G) = X \times Y$, where $X = \{1, 2, \ldots, K\}$ denotes the K channels, and $Y = \{1, 2, \ldots, M\}$ denotes the M SUs. For any edge $(k, m) \in E(G)$, the weight $w_{k,m}$ is given by the secondary throughput of user m on channel k, which is a function of the transmit power P_k and the sensing threshold ϵ_k. Note that in (3.34) and (3.35), the miss detection probability $P_{m,k}$ is increasing with ϵ_k, while the false alarm probability $P_{f,k}$ is decreasing with ϵ_k. Thus, equation $P_{m,k}(P_k, \epsilon_k^*) = P_m$ decides the optimal sensing threshold ϵ_k^*, which is given by:

$$\epsilon_k^* = \left(\frac{Q^{-1}(1 - P_m)}{\sqrt{N_s}} + 1 \right) (1 + \gamma_{s,k} + \gamma_{i,k}). \tag{3.38}$$

Thus, the weight of edge (k, m) can be written as:

$$w_{k,m}(P_k) = \beta_k \left[1 - P_{f,k}(P_k, \epsilon_k^*) \right] \log \left(1 + \frac{P_k |h_{k,m}|^2}{\sigma_u^2} \right). \tag{3.39}$$

Note that in (3.39), the multiplier $[1 - P_{f,k}(P_k, \epsilon_k^*)]$, which represents the probability that the spectrum hole is correctly detected, is decreasing with the transmit power P_k, while the multiplier $\log(1 + P_k |h_{k,m}|^2 / \sigma_u^2)$, which represents the channel capacity, is increasing with P_k. Therefore, we can expect the total throughput first increases and then decreases with P_k, and we denote by $P_{k,m}^{\max}$ as the turning point. The value of $P_{k,m}^{\max}$ can be numerically calculated by solving $\partial w_{k,m} / \partial P_k = 0$.

To solve the above two-dimensional matching problem, we first present some basic concepts. A *feasible labeling* l is a mapping from each vertex $v \in V(G)$ to $\mathbb{R}^+$, where $l(x) + l(y) \geq w_{x,y}$. And an *l-equal* problem is a two-dimensional matching problem $G_l = (X, Y)$ where $V(G_l) = V(G), E(G_l) = \{(x, y) | l(x) + l(y) = w_{x,y}\}$ and $w_{x,y}^l \equiv 1$. Briefly, the l-equal problem considers a subgraph G_l of the original problem, and it aims to find the matching with the largest size. If the optimal matching S_l^* of the l-equal problem covers all vertexes in G, for which we call S_l^* a *perfect matching* for G, we have

$$w(S_l^*) = \sum_{(x,y) \in S_l^*} w_{x,y} = \sum_{v \in V(G)} l(v). \tag{3.40}$$

On the other hand, for any matching S of G, we have

$$w(S) = \sum_{(x,y) \in S} w_{x,y} \leq \sum_{v \in V(G)} l(v). \tag{3.41}$$

Therefore, we have $w(S_l^*) \geq w(S)$, that is, $S_l^* = S^*$ is the maximum weighted matching of G. Now, the two-dimensional matching problem can be seen as two separate problems. The first one is to solve the l-equal problem for a feasible labeling l, and the second one is to find a feasible labeling l such that the solution S_l^* of the

Table 3.3 Hungarian algorithm

Input: $G_l = (X, Y)$

Output: the largest matching S_l^*

1: Set $S = \emptyset$ as the initial matching, and set $A = X$ as the vertexes in X that are not matched by S.

2: If $A = \emptyset$, stop and output $S_l^* = S$; If $A \neq \emptyset$, take $x \in A$, set $U = \{x\}, V = \emptyset$, and turn to the next step.

3: Set $N(U) = \{y \in Y | \exists x \in U, (x, y) \in E(G_l)\}$ as the neighbors of U. If $N(U) \subseteq V$, there is no augmenting path for S that starts from node x. Then set $A = A - \{x\}$ and turn to step 2; If $N(U) \nsubseteq V$, take $y \in N(U) - V$ and turn to the next step;

4: If $\exists z \in X, (z, y) \in S$, then set $U = U \cup \{z\}$ and $V = V \cup \{y\}$, and turn to step 3; If $\nexists z \in X, (z, y) \in S$, then we have an augmenting path P with the starting point x and the end point y. Set $S = S \oplus E(P)$ and $A = A - \{x\}$ and turn to step 2.

---

l-equal problem is a perfect matching for G. These two subproblems can be solved by the *Hungarian algorithm* and *label amendment algorithm*, respectively.

Hungarian Algorithm The Hungarian algorithm is to find the largest matching of a bipartite graph G_l for an l-equal problem. For any initial matching $S \in E(G_l)$, the algorithm iteratively increases the size of S by finding an *augmenting path*, which is defined as a path $P = (v_1, v_2, \ldots, v_L)$ where the edges $(v_1, v_2), (v_2, v_3), \ldots, (v_{L-1}, v_L)$ belong alternatively to $E(G_l) \backslash S$ and S, and the starting point $v_1 \in X$ and the end point $v_L \in Y$ are not matched by S, i.e., $\nexists (x, y) \in S$ such that $x = v_1$ or $y = v_L$. Note that the edges that do not belong to S are one more than the edges that belong to S, i.e., $|E(P) \cap (E(G_l) \backslash S)| - |E(P) \cap S| = 1$. Therefore, we can get a larger matching $S \oplus E(P)$ by replacing the edges in $E(P) \cap S$ with the edges in $E(P) \cap (E(G_l) \backslash S)$. The algorithm stops when there is no augmenting path for the current matching and the output is the largest matching S_l^*. We formally present the Hungarian algorithm in Table 3.3.

Label Amendment Algorithm To ensure the existence of a perfect matching, some vertexes and zero-weight edges are added to $G = (X, Y)$ to make it a complete bipartite graph where $|X| = |Y|$ and $E(G) = \{(x, y) | x \in X, y \in Y\}$. For any initial labeling l, we use the Hungarian algorithm to calculate the largest matching S_l^* for the corresponding l-equal problem. If S_l^* is a perfect matching for G, then output $S^* = S_l^*$ as the maximum weighted matching for the original two-dimensional problem. Otherwise, the Hungarian algorithm outputs $U \subset X, V \subset Y$, and we have $\{(x, y) | x \in U, y \in Y - V\} \notin E(G_l)$. Let $\theta_l = \min\{l(x) + l(y) - w_{x,y} | x \in U, y \in Y - V\}$ denote the "extra" labeling for these edges, and we amend the vertex labeling as follows:

$$l'(v) = \begin{cases} l(v) - \theta_l, & v \in U, \\ l(v) + \theta_l, & v \in V, \\ l(v), & \text{others.} \end{cases} \tag{3.42}$$

Table 3.4 Label amendment algorithm

Input: $G = (X, Y)$ and $\{w_{x,y}\}$

Output: the maximum weighted matching S^*

1: Add vertexes and zero-weight edges to make $G = (X, Y)$ a complete bipartite graph with $|X| = |Y|$.
2: Set an initial labeling $l(x) = \max_{y \in Y} w_{x,y}$ for all $x \in X$, and $l(y) = 0$ for all $y \in Y$.
3: Find the largest matching S_l^* of the l-equal problem G_l using the algorithm in Table 3.3.
4: If S_l^* is a perfect matching for G, then output $S^* = S_l^*$; Otherwise, set the vertex labeling l' as in (3.42) and turn to step 3.

We can verify that l' is still a feasible labeling, and the edges $\{(x,y)|x \in U, y \in Y - V\}$ will be included in the l'-equal problem $G_{l'}$. Then, we find the largest matching of the l'-equal problem. The algorithm stops when it outputs a perfect matching for G, and we formally present the label amendment algorithm in Table 3.4.

Proposed Joint Spectrum Access and Power Allocation Algorithm The above methods can find the optimal spectrum access strategy for a given power allocation. In our problem, we iteratively allocate the power units to the channels, and in iteration t, we assume the power allocation is given by $\{P_k(t)\}$ and the corresponding optimal spectrum access strategy is given by matching $S^*(t)$ with the final vertex labeling l_t. Thus, we have $P_{k^*}(t+1) = P_{k^*}(t) + \Delta$, where

$$k^* = \arg \max_{k: \exists m, (k,m) \in S^*(t)} \{w_{k,m}(P_k(t) + \Delta) - w_{k,m}(P_k(t))\} \qquad (3.43)$$

is the channel with the largest marginal rate, and $P_k(t+1) = P_k(t)$ for the rest channels. And the initial feasible labeling of the next iteration $t+1$ is given by

$$l(v) = \begin{cases} l_t(k^*) + \pi_t, & v = k^*, \\ l_t(v), & \text{others,} \end{cases} \qquad (3.44)$$

where

$$\pi_t = \max_{m=1,2,\dots,M} \{w_{k^*,m}(P_{k^*}(t+1)) - w_{k^*,m}(P_{k^*}(t))\}. \qquad (3.45)$$

The algorithm stops when all power units are allocated, or at an iteration t such that

$$P_k(t) = P_{k,m}^{max}, k = 1, 2, \dots, K, \qquad (3.46)$$

where the channel-user pair (k, m) is the matched in $S^*(t)$. We formally present our algorithm in Table 3.5.

Table 3.5 Joint spectrum access and power allocation algorithm

---

1: Set $P_k = 0$ for all $k = 1, 2, \ldots, K$ and the initial spectrum access strategy $S^*(0) = \emptyset$.

2: **while** $t < M$ and (3.46) are not satisfied. **do**

3:    Choose channel k^* as in (3.43), and add one power unit Δ to this channel.

4:    Set the labeling l as in (3.44), and calculate the optimal spectrum access strategy $S^*(t+1)$ using the algorithm given in Table 3.4.

5:    Set $t \rightarrow t + 1$;

6: **end while**

7: The final power allocation is $\{P_k(t)\}$ and the final spectrum access is $\{\alpha_{k,m}\}$ where $\alpha_{k,m} = 1$ if and only if $(k, m) \in S^*(t)$.

Table 3.6 Parameters for simulation

Parameters	Value
The probability of the PU's transmission P_{on}	0.5
The miss detection constraint P_m	0.1
The received SNR of the PU signal $\gamma_{s,k}$	$-10\,\mathrm{dB}$
The number of samples in a sensing slot N_s	200
The ratio of the transmit power and noise P	100–120 dB
The ratio of the interference and the power χ^2	-110 to $-90\,\mathrm{dB}$
The number of subcarriers K	20
The number of SUs M	10–20
The upper bound of iteration number I	10–50

3.2.2.3 Results

In this part, we present the performance of the proposed matching algorithm, compared with the random algorithm and the greedy algorithm. In the random algorithm, the SBS randomly chooses an SU for each idle channel, while in the greedy algorithm, the SBS sequentially chooses the optimal SU for each idle channel. The parameters of our simulations are given in Table 3.6.

In Fig. 3.11, we show the throughput of the SBS as a function of the number of SUs M for each algorithm. Due to the multi-user diversity, the secondary throughput of the SBS increases with the number of SUs. Compared with the greedy algorithm and the random algorithm, our proposed matching algorithm increases the throughput of the SBS by 20 and 140 %, respectively, when the number of SUs $M = K = 10$. When the number of SUs increases, the probability that the channels have different optimal SUs also increases. Therefore, we can expect the greedy algorithm also achieves a high multi-user diversity, and its performance can approach the proposed matching algorithm, as seen in Fig. 3.11 when $M = 20$.

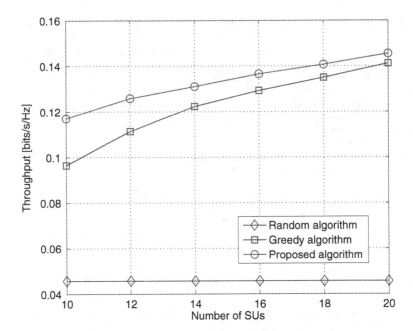

Fig. 3.11 Throughput of the SBS as a function of the number of SUs with $P = 100\,\text{dB}$ and $\chi^2 = -110\,\text{dB}$

In Fig. 3.12, we show the throughput of the SBS as a function of the transmit power P for each algorithm. In the greedy algorithm and the random algorithm, the throughput of the SBS first increases and then decreases with the transmit power. The reason is that the increasing transmit power also increases the self-interference at the sensing antenna, which degrades the sensing performance of the SBS and leads to the decrease in overall throughput, as seen in (3.39). However, in our proposed algorithm, the transmit power is gradually allocated to the optimal channel-user pair. If all channel-user pairs have reached their maximal transmit power, as seen in (3.46), our algorithm will stop and drop the rest power. Therefore, throughput of the SBS does not decrease when the transmit power is high, as seen in Fig. 3.12 when $P > 117\,\text{dB}$.

In Fig. 3.13, we show the throughput of the SBS as a function of the SIS capability χ^2 for each algorithm. When the SIS capability of the SBS declines, the self-interference at the sensing antenna is increased. In order to ensure the miss detection constraint in (3.36b), the optimal sensing threshold (3.38) should be set a low value, which results in a high probability of false alarm and leads to the waste of a lot of transmission opportunities. Therefore, no matter what algorithm we use, the throughput of the SBS always decreases as χ^2 increases, as we can see in Fig. 3.13.

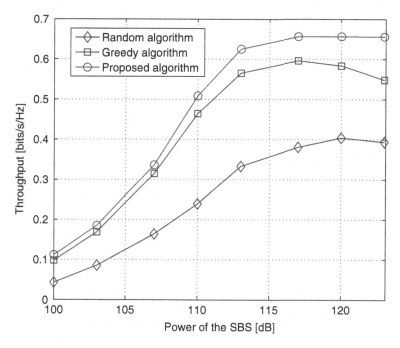

Fig. 3.12 Throughput of the SBS as a function of transmit power P with $M = 10$ and $\chi^2 = -107$ dB

In Fig. 3.14, we show the tradeoff between the performance and the computation complexity of the proposed algorithm. In our proposed algorithm, the total power is equally divided into I power units, and the number of iterations is bounded by the value of I. When I is increased, the computation complexity increases, while at the same time, the size of a power unit $\Delta = P/I$ decreases, which makes it possible to get a power allocation that is closer to the optimal solution. Thus, the performance of our proposed algorithm increases with the number of iterations, as seen in Fig. 3.14.

It can be seen from the discussion of this part that when we introduce the full-duplex technology in a cognitive cellular network, the SBS can simultaneously sense the spectrum and transmit to the SUs. In such a network, the joint spectrum access and power allocation problem has been formulated as a three-dimensional matching problem, and we have proposed an approximate solution based on a two-dimensional matching algorithm. Both the analysis and simulations show that the secondary throughput of the SBS does not monotonically increase with the total transmit power, and there exists an upper bound when the transmit power is sufficiently high. Also, the secondary throughput is highly influenced by the number of SUs and the SIS capability. Compared with the greedy algorithm and the random algorithm, our proposed algorithm can highly increase the secondary throughput of the SBS in various conditions.

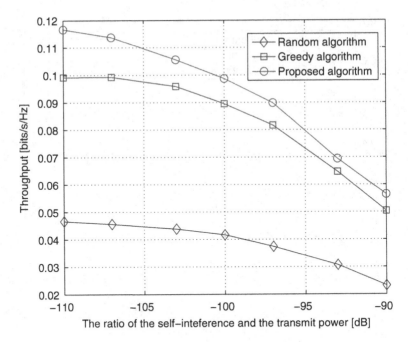

Fig. 3.13 Throughput of the SBS as a function of the self-interference suppression capability χ^2 with $M = 10\,\text{dB}$ and $P = 100\,\text{dB}$

3.3 Conclusion

In this chapter, we studied the networks with multiple FD cognitive users. Both cooperation and contention among users have been discussed. In Sect. 3.1, we provided and analyzed a feasible cooperative spectrum sensing scheme based on the LAT protocol that allows SUs to perform cooperative spectrum sensing and access to the spectrum simultaneously, in which the interference from the transmitting SU to other cooperative SUs becomes a major problem. A comparison between the cooperative LAT and conventional cooperative LBT protocols has shown the improvement of secondary throughput under low sensing SNR. In Sect. 3.2, both the distributed and centralized scenarios were considered. For the distributed scenario, a decentralized DSA protocol based on the CSMA scheme was presented, and proved effective in the simulation part. For the centralized case, a cognitive cellular network was considered. Specifically, we studied the joint spectrum and power allocation problem in it and proposed a sub-optimal allocation algorithm based on two-dimensional matching theories and algorithms.

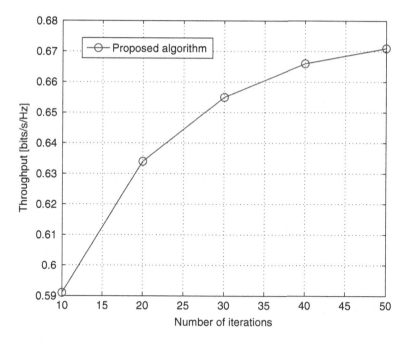

Fig. 3.14 Throughput of the SBS as a function of the power unit size Δ with $M = 10, P = 100\,\text{dB}$ and $\chi^2 = -107\,\text{dB}$

References

1. Y. Liao, T. Wang, L. Song, and B. Jiao, "Cooperative Spectrum Sensing for Full-duplex Cognitive Radio Networks," in *Proc. IEEE Global Communication Conferences (ICCS)*, Macau, China, Nov. 2014.
2. Y. Liao, T. Wang, K. Bian, L. Song, and Z. Han, "Decentralized Dynamic Spectrum Access in Full-Duplex Cognitive Radio Networks," in *Proc. IEEE International Conference on Communications (ICC)*, London, UK, Jun. 2015.
3. T. Wang, Y. Liao, B. Zhang, and L. Song, "Joint Spectrum Access and Power Allocation in Full-Duplex Cognitive Cellular Networks," in *Proc. IEEE International Conference on Communications (ICC)*, London, UK, Jun. 2015.
4. S. Huang, X. Liu, and Z. Ding, "Opportunistic Spectrum Access in Cognitive Radio Networks," in *IEEE InfoCom 2008*, Phoenix, AZ, Apr. 2008.
5. H. Kim and K. G. Shin, "Efficient Discovery of Spectrum Opportunities with Mac-Layer Sensing in Cognitive Radio Networks," *IEEE Transactions on Mobile Computing*, vol. 7, no. 5, pp. 533–545, May 2008.
6. A. Motamedi and A. Bahai, "Mac Protocol Design for Spectrum-Agile Wireless Networks: Stochastic Control Approach," in *Proc. IEEE International Symposium on New Frontiers in Dynamic Spectrum Access Networks (DySPAN)*, Dublin, Ireland, Apr. 2007.
7. M. R. Garey and D. S. Johnson, *Computers and Intractability: A Guide to the Theory of NP-Completeness*, San Francisco, CA: W. H. Freeman and Company, 1979.
8. H. W. Kuhn, "The Hungarian Method for the Assignment Problem," *Naval Research Logistics Quarterly*, vol. 2, pp. 83–97, 1955.
9. J. E. Hopcroft, R. M. Karp, "An $n^{5/2}$ Algorithm for Maximum Matchings in Bipartite Graphs," *SIAM Journal on Computing*, vol. 1, no. 4, pp. 225–231, 1973.

Chapter 4
Full-Duplex WiFi

WiFi technologies have received a rapid proliferation over the past decade [1]. In particular, for conventional HD WiFi networks, the CSMA/CA protocol is adopted in MAC layer; and carrier sensing is performed in the PHY layer to detect the channel state

In particular, for traditional WiFi networks, the CSMA/CA protocol is adopted in MAC layer [2–5], in which carrier sensing and data transmission are divided in time domain. In this method, once collision happens between some users, they need to finish the whole collided packet and wait for the absence of the ACK signal to be aware of the collision. Thus, conventional HD devices in WiFi networks always suffer from long collision, and this becomes a severe problem when the network is crowded, and collision happens frequently.

The development of self-interference cancellation and suppression techniques [6] have made the FD communications possible. With the help of FD techniques, some FD-MAC protocols have been proposed recently [6–9]. In [6] and [7], the centralized FD-MAC protocols are proposed: [6] considers bidirectional transmission between a pair of nodes, and uses busytone to eliminate the hidden terminal problem, and [7] design the protocol with three new elements, namely, shared random backoff, header snooping and virtual backoffs. Decentralized FD-MAC protocols are proposed in [8, 9] based on CSMA/CA. The former mainly focuses on bidirectional transmission between users, and the latter discusses simultaneous transmissions among two or three FD users. However, the way how to fully utilize FD techniques for WiFi networks with multiple contending users, and comprehensive analysis from PHY to MAC layers still requires further investigation.

To this end, in this chapter, we present a new MAC protocol for FD WiFi networks in which a FD users compete for transmission opportunities over a single channel [10]. Due to FD techniques, users are able to sense and monitor the channel usage state while they are transmitting. Thus, if collision happens, users are able to backoff in short time before finishing the whole data packet and waiting for the absence of the ACK packet. Note that different from the CSMA/CD in Ethernet,

© The Author(s) 2016
Y. Liao et al., *Listen and Talk*, SpringerBriefs in Electrical and Computer
Engineering, DOI 10.1007/978-3-319-33979-5_4

FD-WiFi operates in wireless environment with fading channels and RSI caused by FD techniques. Thus, the sensing performance becomes one of the important issues that require careful study. In the rest of this chapter, we present the design of a cross-layer FD WiFi protocol, and derive the analytical throughput of the proposed FD-WiFi protocol by taking imperfect sensing caused by RSI into consideration. Simulation results are provided to prove the effectiveness of the proposed protocol.

4.1 System Model

We consider a WiFi network consisting of one access point (AP) and M FD-enabled users $\{U_1, \ldots, U_M\}$, where the users are independently and randomly distributed in the coverage area of the AP. We focus on the uplink traffic, in which data packets are transmitted from the users to the AP by our proposed FD-WiFi protocol, and each user is assumed to always have a packet to transmit with the same transmission power. The channel can serve at most one user at a time, otherwise collision happens. Therefore, each user, equipped with two antennas, performs carrier sensing to detect the channel state and contends for the idle channel against each other by the proposed protocol. When a certain user, say U_m ($m \in \{1, 2, \ldots, M\}$), accesses the channel, it uses one antenna for carrier sensing and the other antenna for data transmission simultaneously. However, the self-interference between those two antennas leads to imperfect sensing. Therefore, carrier sensing errors need to be considered to evaluate the performance of FD-WiFi protocol.

4.2 Cross-Layer Protocol Design

In this section, we propose our cross-layer protocol design, over both the PHY layer and MAC layer, for FD-WiFi networks. We first discuss the FD carrier sensing with sensing errors, then the FD-WiFi protocol is proposed based on the sensing performance in the PHY layer and the new adaption to realize simultaneous carrier sensing and data transmission in the MAC layer.

4.2.1 Full-Duplex Carrier Sensing

In the analysis of conventional HD-WiFi networks [2], noise is often neglected and sensing is assumed perfect. For simplicity and comparison fairness, we also omit the noise in this letter. Thus, a user has a perfect sensing without transmitting. And we only need to analyze the imperfect sensing due to RSI for a transmitting user.

Furthermore, when the transmitting user collides with more than one users, the accumulated collision signal usually overwhelms RSI, and we assume perfect sensing in this situation. Thus, we only consider the sensing errors in the following two cases: (1) $\mathcal{H}_0$: the transmitting user singly occupies the channel; (2) $\mathcal{H}_1$: the transmitting user has a collision with another user. When the channel is occupied, the sensing signal at the transmitting user can be given by

$$
y = \begin{cases} h_r s_t, & \mathcal{H}_0, \\ h_r s_t + h_c s_c, & \mathcal{H}_1, \end{cases}
\tag{4.1}
$$

where s_t denotes the signal of transmitting user and s_c is the signal of collided user, both of which have the same transmission power, and h_r and h_c denote corresponding self-interference channel and collision channel, respectively. We adopt typical path loss Rayleigh fading model for all channels. Therefore, $h_c s_c$ is zero-mean complex Gaussian distributed with average power $\overline{P}_r(\frac{\overline{d}}{d})^\alpha$, where α is the path loss exponent and $\overline{P}_r$ is the reference received signal power at the reference distance $\overline{d}$ (usually 1 m), and d is the distance between the transmitting user and the collided user. Moreover, according to [6], $h_r s_t$ is also a complex Gaussian variable with zero mean and average power $\chi^2 \overline{P}_r$), where χ is SIS factor.

As for the sensing strategy, energy detection is adopted, and we assume the process is time-slotted, thus the sensing test statistics can be given by

$$
O = \frac{1}{N_s} \sum_{n=1}^{N_s} |y(n)|^2,
\tag{4.2}
$$

where $y(n)$ denotes the nth sample of sensing signal, and N_s is the sampling number in one slot.

The transmitting user compares O with the carrier sensing threshold to decide whether a collision happens or not. And two types of sensing errors exist, namely false alarm and miss detection, respectively. Specifically, false alarm wastes available channel slots, while miss detection causes collisions. And we need to balance a tradeoff between them: a higher sensing threshold decreases the false alarm probability while it increases the miss detection probability.

4.2.2 FD-WiFi MAC Protocol

Figure 4.1 shows the proposed MAC protocol for FD-WiFi, which consists of the following several parts.

Sensing All users keep sensing the channel continuously regardless of its own activity, and make decisions of the channel usage at the end of each slot with duration τ, in which N_s samples are received by each user.

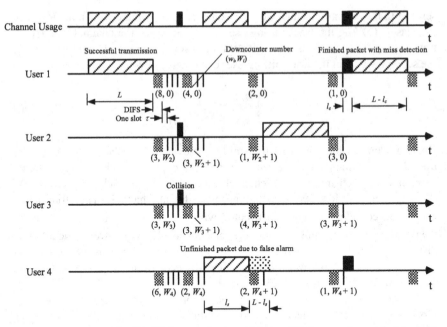

Fig. 4.1 MAC Protocol for FD-WiFi networks, in which (w_i, W_i) denotes the residual backoff time and the backoff stage of user i

Backoff Mechanism Once the channel is judged idle without interruption for a certain period of time as long as a distributed interference space (DIFS) (shown as the dotted area below each line), users check their own backoff timers and generate a random backoff time for additional deferral if their timers have counted down to zero. The additional backoff time after a DIFS is also slotted by τ, i.e., the backoff time is expressed as

$$\text{Backoff Time} = w \times \tau = \text{Random (CW)} \times \tau, \qquad (4.3)$$

where $\text{CW} = 2^W \cdot \text{CW}_{\min}$ is the contention window length, and $w = \text{Rand (CW)}$ is a random integer drawn from the uniform distribution over the interval $[0, \text{CW})$, where $W \in [0, W_{\max}]$ is the backoff stage depending on the number of unsuccessful transmissions for a packet. The countdown starts right after the DIFS, and suspends when the channel is detected occupied by others.

Channel Access and Transmission Suspension A user accesses the channel and begins transmission when its timer reaches zero. During the transmission, if it detects the signal from other users, it stops its transmission and switches to the backoff procedure immediately. If the packet is finished, the user resets the backoff state $W = 0$. Otherwise, it sets $W = \min \{W + 1, W_{\max}\}$.

4.3 Performance Analysis

In this section, we analyze the performance of the proposed protocol. We first discuss false alarm and miss detection for carrier sensing, then we derive the normalized throughput of FD-WiFi networks.

4.3.1 Imperfect Carrier Sensing

In this section, we mainly derive the expressions of false alarm probability (p_f) and miss detection probability (p_m) for the sensing performance.

As shown in (4.2), O is the sum of sampling signal power in one slot, and thus according to [11], O is Gamma distributed, the probability density function of which can be expressed as

$$f_O(x) = \frac{x^{N_s - 1} e^{-x/\phi}}{\phi^{N_s} \Gamma(N_s)}, \tag{4.4}$$

where $\phi = \chi^2 \overline{P}_r$, and $\phi = \left(\chi^2 + \left(\frac{\overline{d}}{d}\right)^\alpha\right)\overline{P}_r$ for two cases $\mathcal{H}_0$ and $\mathcal{H}_1$, respectively.

Therefore, for a certain sensing threshold ϵ, we can get the expressions of p_f and p_m, respectively,

$$p_f = \mathrm{P_r}(O > \epsilon \,|\, \mathcal{H}_0) = 1 - \Gamma\left(N_s, \frac{\epsilon}{\chi^2 \overline{P}_r}\right),$$

$$p_m(d) = \mathrm{P_r}(O < \epsilon \,|\, \mathcal{H}_1) = \Gamma\left(N_s, \frac{\epsilon}{\left(\chi^2 + \left(\frac{\overline{d}}{d}\right)^\alpha\right)\overline{P}_r}\right), \tag{4.5}$$

where $\Gamma(m, x) = \frac{1}{\Gamma(m)}\int_0^x t^{m-1} e^{-t}\,dt$ is the incomplete gamma function. Note that $p_m(d)$ is related to the transmitting user and the collided users, while p_f is not. We can rewrite the expressions of ϵ and $p_m(d)$ as the following:

$$\epsilon = a\chi^2 \overline{P}_r,$$

$$p_m(d) = \Gamma\left(N_s, a - \frac{b}{d^\alpha + c}\right), \tag{4.6}$$

where $a = \Gamma^{-1}(N_s, 1 - p_f)$, $b = a\frac{\overline{d}^\alpha}{\chi^2}$, and $c = \frac{\overline{d}^\alpha}{\chi^2}$.

Furthermore, users are independently and randomly distributed in the coverage area of AP, the radius of which is denoted by R. Then we can get the average miss detection probability

$$p_m = \frac{2}{\pi R^4} \int_0^R \int_0^R \int_0^{2\pi} \Gamma\left(N_s, a - \frac{b}{d^\alpha + c}\right) r_1 r_2 d\theta dr_1 dr_2, \qquad (4.7)$$

where r_1, r_2 are the distances of transmitting user and collided user away from AP, θ is their included angle, and $d = \sqrt{r_1^2 + r_2^2 - 2r_1r_2\cos\theta}$ is the distance between them. We can find that the expression of p_m is related to the path loss exponent. Particularly, when free space channel is considered, i.e., $\alpha = 2$, we can get an approximation of p_m as

$$p_m \approx \frac{2}{3}\Gamma\left(a - \frac{b}{R^2 + c}\right) + \frac{1}{6}\Gamma\left(N_s, a - \frac{b}{2R^2 + c}\right). \qquad (4.8)$$

Since the average power of self-interference is positively proportional to χ^2, a higher SIS factor leads to worse sensing performance, which is consistent with (4.8). Furthermore, according to (4.5)–(4.8), p_f is negatively related to ϵ, while p_m is positively related. Therefore, there exists a tradeoff between the false alarm probability and miss detection probability.

4.3.2 Spectrum Utilization Efficiency and Throughput

In this section, we bring the sensing performance in PHY layer to MAC layer, and study the analytical performance of the proposed MAC protocol for FD-WiFi. Note that when only one user is transmitting, all other users can detect its transmission perfectly, which means that once a collision-free transmission begins, it either completes the packet or suspends it because of false alarm. This process is independent with other users' sensing and contending, and thus, contention and transmission can be considered separately. We first derive the probability that one transmission attempt collides with other transmissions. Then by considering the average successful transmission length, we evaluate the saturation throughput.

4.3.2.1 Collision Probability

We follow the assumption in [2] that each packet gets collided with a same probability independent of the value of CW_i. Let $\{w_i, W_i\}$ denote the state of the ith contending user. For each user, the state change can be modeled as a discrete-time Markov chain illustrated in Fig. 4.2 [12]. The non-zero transition probabilities are given as

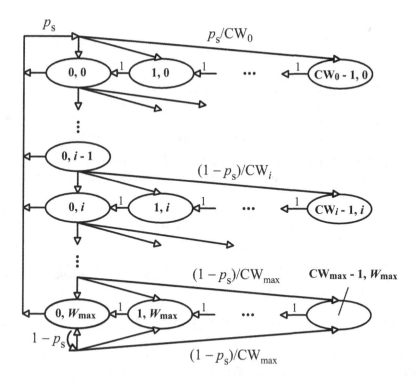

Fig. 4.2 Markov chain of the backoff window size

$$
\begin{cases}
P\left(w_i - 1, W_i | w_i, W_i\right) = 1, & w_i \in (0, \mathrm{CW}_i)\,, W_i = [0, W_{\max}]\,, \\
P\left(w_i, 0 | 0, W_i\right) = p_s / \mathrm{CW}_{\min}, & w_i \in [0, \mathrm{CW}_{\min})\,, W_i = [0, W_{\max}]\,, \\
P\left(w_i, W_i + 1 | 0, W_i\right) = (1 - p_s) / \mathrm{CW}_{i+1}, & w_i \in [0, \mathrm{CW}_{i+1})\,, W_i = [0, W_{\max})\,, \\
& w_i \in [0, \mathrm{CW}_{\max})\,,
\end{cases}
$$

$$(4.9)$$

where p_s denotes the probability that the considered user successfully finishes its transmission without awareness of collision. Note that p_s does not equal to the non-collision probability due to imperfect sensing. Specifically, if two users collide, it is possible that only one user stops, and when one user is transmitting without collision, it may cease the transmission due to false alarm.

Consider the steady-state distribution of the Markov chain, the probability that one user stays in each state can be calculated. Let $p_{w,W}$ denote the probability that one user is in the state of $\{w, W\}$, and the probability that a certain user begins transmission in the next slot is

$$p = \sum_{W=0}^{W=W_{\max}} p_{0,W}$$

$$= \frac{2(2p_s - 1)}{(2p_s - 1)(\mathrm{CW}_{\min} + 1) + (1 - p_s)\mathrm{CW}_{\min}\left(1 - (2 - 2p_s)^{W_{\max}}\right)}. \tag{4.10}$$

Then, consider the relation between p_s and p. For simplicity, we assume the packet length L is fixed. The calculation of p_s has two pre-requisites:

1. The probability that one user starts collision-free transmission after colliding with others for l slots, denoted as $p_a(l), l \in [0, L]$, which can be expressed as

$$p_a(l) = \begin{cases} (1-p)^{M-1} & l = 0, \\ (M-1)p(1-p)^{M-2}p_m^{2l-1}(1-p_m) & 1 \le l \le L-1, \\ (M-1)p(1-p)^{M-2}p_m^{2L-1} & l = L. \end{cases} \tag{4.11}$$

2. The probability of successfully finishing current packet with residual collision-free length of l, denoted as $p_b(l), l \in [0, L]$:

$$p_b(l) = \left(1 - p_f\right)^l \quad 0 \le l \le L. \tag{4.12}$$

Successful transmission requires at least one user transmits the entire packet without the awareness of collision. Thus, p_s can be calculated as

$$p_s = \sum_{l=0}^{L} p_a(l)\, p_b(L-l)$$

$$= (1-p)^{M-1}\left(1-p_f\right)^L + (M-1)p(1-p)^{M-2}p_m \cdot \frac{\left(1-p_f\right)^L - p_m^{2L}}{1 - p_f - p_m^2}. \tag{4.13}$$

Combining (4.10) and (4.13), the values of p and p_s can be solved numerically.

4.3.2.2 Throughput

We use the time fraction that the channel is occupied for successful transmission as the normalized throughput, i.e., the throughput is defined as,

$$C = \frac{\mathbb{E}\,[\text{Successful transmission length}]}{\mathbb{E}\,[\text{Consumed time for a successful transmission}]}$$

$$= \frac{P_s L_s}{P_e + P_s\,(L_s + \mathrm{DIFS}) + P_c\,(L_c + \mathrm{DIFS})}, \tag{4.14}$$

where $P_s = Mp(1-p)^{M-1}$ denotes the probability that a successful transmission occurs, $P_e = (1-p)^M$ is the probability that the channel is empty, $P_c = 1 - P_e - P_s$ represents the collision probability, and L_s, L_e, and L_c denote the average length of successful transmission, empty state, and collision, respectively. The average length of successful transmission and collision can be calculated as, respectively,

$$
\begin{aligned}
L_s &= \sum_{l=1}^{L-1} l(1-p_f)^{l-1} p_f + L(1-p_f)^{L-1} \\
&= \frac{1-(1-p_f)^{L-1}}{p_f} + (1-p_f)^{L-1},
\end{aligned}
\tag{4.15}
$$

$$
\begin{aligned}
L_c &= \left(P_c + \binom{M}{2} p^2(1-p)^{M-2} \sum_{l=1}^{L-1} p_m^{2l}(1-p_m^2)\, l \right) / P_c \\
&= 1 + \binom{M}{2} p^2(1-p)^{M-2} \frac{p_m^2(1-p_m^{2L-2})}{P_c(1-p_m^2)}.
\end{aligned}
\tag{4.16}
$$

The throughput is readily obtained by substituting (4.15) and (4.16) into (4.14).

4.3.3 Comparison with the Basic CSMA/CA Mechanism

We make a comparison between the proposed MAC protocol for FD-WiFi with the conventional CSMA/CA in this section. For fairness, we consider the same system with M users, and omit the noise term. The analytical performance of CSMA/CA is elaborated in [2], which are omitted here due to the space limitation. Some main differences between the two protocols are listed as follow.

- **Collision length.** In conventional CSMA/CA, the "blindness" in transmission results in long collision, which is typically a packet length. FD allows users to detect collision while transmitting. Thus, the average collision length L_c, as is derived in (4.16), is slightly more than one slot, which is sharply reduced compared with CSMA/CA.
- **Successful transmission length.** In CSMA, once a collision-free transmission begins, it can always be finished successfully without interruption. However, in FD-WiFi, the transmission may get ceased due to false alarm, especially for long packets. According to (4.15), if L is sufficiently large, L_s goes to $1/p_f$. Also, false alarm leads to unnecessary backoff and increase of contention window, which may further degrade the performance of FD-WiFi. Thus, in FD-WiFi, the design of an appropriate packet length should be carefully considered.

4.4   Simulation Results

In this section, simulation results are given to show the throughput performance of FD-WiFi in terms of different parameters. We consider 20 contending users uniformly distributed in the coverage area of AP, whose radius R is set as 10 m. For the fading channel, we set $\alpha = 2$ and $\overline{P}_r = 10$ mW with $\overline{d} = 1$ m. The slot sampling number N_s is set as 100. Furthermore, $CW_{\min}$ and $CW_{\max}$ are set to be 2^3 and 2^8 slots, respectively. Finally, packet length is fixed to be 100 slots and DIFS is 2 slots. We run for 10^6 transmission attempts to get the simulated results.

Figure 4.3 shows the throughput versus false alarm probability. Both simulated (the pentagrams) and analytical (corresponding lines) results of FD-WiFi protocol are presented, and the throughput of traditional HD-WiFi with CSMA/CA protocol in MAC layer is also provided to show the improvement of FD-WiFi. Two different lines are given for FD-WiFi, with $\chi^2 = 0.15$ and 0.3, respectively. Since conventional HD-WiFi has perfect sensing, its throughput is stable, while the throughput of FD-WiFi varies with p_f. Figure 4.3 shows that throughput gets considerably improved in FD-WiFi and there exists an optimal value of p_f to achieve the maximum throughput. A smaller p_f leads to larger p_m and overlong collision length, while a larger p_f makes the average successful transmission length too short. Since p_f is determined by ϵ, the carrier sensing threshold should be well-designed to achieve the maximum throughput. We can also find that the optimal p_f with a

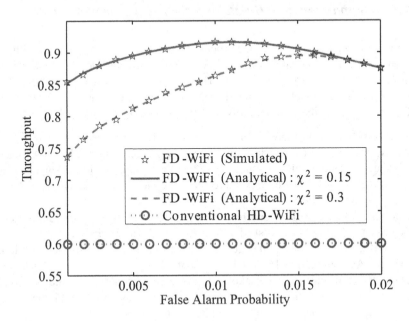

Fig. 4.3 Throughput versus false alarm probability

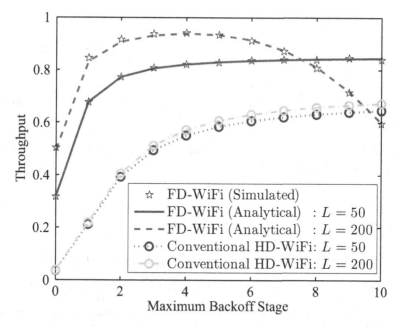

Fig. 4.4 Throughput versus maximum backoff stage

lower SIS factor is smaller than that with a higher SIS factor. The reason is that a higher χ^2 means more severe RSI, then p_m is also larger with a certain p_f, and thus p_f should increase to achieve the maximum throughput.

In Fig. 4.4, we show the relationship between the normalized throughput and maximum backoff stage $W_{\max}$, with $CW_{\min} = 2^3$, $\chi^2 = 0.15$ and p_f fixed to be 0.01. There exist both simulated and analytical results of FD-WiFi in two cases, with packet length $L = 50$ and 200, respectively. Also the corresponding throughput of conventional HD-WiFi is provided. According to Fig. 4.4, we can also find that the FD-WiFi has a better throughput than the conventional HD-WiFi in most cases. For the conventional HD-WiFi, $CW_{\max}$ increases with $W_{\max}$, and less collision happens; therefore, the throughput increases with $W_{\max}$ and becomes stable with enough high $W_{\max}$. While for the FD-WiFi, when $W_{\max}$ is small, collision is considerably severe, and thus the throughput also increases with $W_{\max}$. However, different from the conventional HD-WiFi, the throughput of FD-WiFi may drop with enough large $W_{\max}$. From Fig. 4.4, the FD-WiFi throughput with $L = 200$ decreases when $W_{\max} > 4$. The reason is that the asymptotic value of L_s is 100 with $p_f = 0.01$, and when packet length is 200 slots, false alarm is quite likely to happen during data transmission and most users choose $CW_{\max}$ as their backoff contention window. Thus, more time is spent in the backoff process and throughput gets smaller, as $W_{\max}$ increases.

4.5 Conclusions

In this chapter, we presented a cross-layer protocol design for FD-WiFi networks to resolve the problem of long collision duration in conventional HD-WiFi networks. By considering the impact of RSI, we analyzed two types of carrier sensing errors, namely false alarm and miss detection, then we combined the sensing performance with the proposed protocol to derive the normalized throughput. We also showed by analysis and simulation results that our proposed protocol design can get improved throughput than conventional HD-WiFi and carrier sensing threshold should be properly designed to achieve the maximum throughput.

References

1. B. P. Crow, I. Widjaja, J. G. Kim, P. T. Sakai, "IEEE 802.11 Wireless Local Area Networks," *IEEE Communications Magazine*, vol. 35, no. 9, pp. 116–126, Sep. 1997.
2. G. Bianchi,"Performance Analysis of the IEEE 802.11 Distributed Coordination Function," *IEEE Trans. on Selected Areas in Comm.*, vol. 18, no. 3, pp. 535–547, Mar. 2000.
3. J. Zhu, X. Guo, L. L. Yang and W. S. Conner, "Leveraging Spatial Reuse in 802.11 Mesh Networks with Enhanced Physical Carrier Sensing," in *IEEE International Conference on Communications*, vol. 7, pp. 4004–4011, Paris, France, Jun. 2004.
4. C. Thorpe, and L. Murphy, "A Survey of Adaptive Carrier Sensing Mechanisms for IEEE 802.11 Wireless Networks," *IEEE Communications Surveys & Tutorials*, vol. 16, no. 3, pp. 1266–1293, Third Quarter 2014.
5. I. Jamil, L. Cariou, J. F. Helard, "Efficient MAC Protocols Optimization for Future High Desity WLANs," in *IEEE Wireless Comms. and Netw. Conf. (WCNC)*, pp. 1054–1059, New Orleans, LA, Mar. 2015.
6. M. Jain, J. I. Choi, T. Kim, D. Bharadia, S. Seth, K. Srinivasan, P. Levis, S. Katti, and P. Sinha, "Practical, Real-time, Full Duplex Wireless," in *Proc. ACM MobiCom 2011*, New York, NY, Sep. 2011.
7. A. Sahai, G. Patel, and A. Sabharwal, Pushing the Limits of Full-Duplex: Design and Real-Time Implementation, Rice University, Houston, TX, USA, Tech. Rep. TREE1104.
8. N. Singh, D. Gunawardena, A. Proutiere, B. Radunovic, H. V. Balan, and P. Key, "Efficient and Fair MAC for Wireless Networks with Self-interference Cancellation," in *Int. Symp. on Modeling and Optimization in Mobile, Ad Hoc and Wireless Networks (WiOpt)*, pp. 94–101, Princeton, NJ, May 2011.
9. S. Goyal, P. Liu, O. Gurbuz, E. Erkip, and S. Panwar, "A Distributed MAC Protocol for Full Duplex Radio," in *Proc. 47th Asilomar Conf. on Signals, Syst. Comput.*, pp. 788–792, Nov. 2013.
10. Y. Liao, K. Bian, L. Song, and Z. Han, "Full-duplex MAC Protocol Design and Analysis," *IEEE Comm. Letters*, vol. 19, no. 7, pp. 1185–1188, Jul. 2015.
11. H. V. Poor, *An introduction to signal detection and estimation*, 2nd ed. New York: Springer-Verlag, 1994.
12. H. Luan, X. Ling, and X. Shen, "MAC in Motion: Impact of Mobility on the MAC of Drive-Thru Internet," *IEEE Trans. on Mobile Computing*, vol. 11, no. 2, pp. 305–319, Feb. 2012.

Chapter 5
Conclusions and Future Works

5.1 Conclusions

In this book, the idea of "listen-and-talk", i.e., enabling simultaneous sensing and transmission in different networks by full-duplex techniques has been presented, and elaborated. Some physical layer characteristics brought by FD techniques have been addressed, and some cross-layer protocol design and resource allocation schemes have been proposed to accommodate with the FD techniques, so that the networks can fully utilize the potential of FD communications. In the FD communication systems, devices can transmit and receive signal with the same time and frequency resource. This attractive feature has the potential to completely change the structure of communication networks, since devices are no longer constrained by the "blindness" during their own transmission. Meanwhile, on the other side, the residual self-interference becomes a key problem in design and implementation of FD networks, which requires careful study in the performance analysis.

We started with the study of full-duplex cognitive radio. In such networks, a SU needs to detect whether there exists a spectrum hole that can be utilized. Different from conventional "listen-before-talk" protocol saying that SUs need to periodically suspend their transmission to perform spectrum sensing, with FD techniques, SUs can simultaneously sense the spectrum and transmit their own packets. We elaborated a "listen-and-talk" (LAT) protocol that allows SUs to sense and transmit on the same channel concurrently, provided detailed performance analysis, and showed its effectiveness by comparing the LAT protocol with conventional listen-before-talk one.

Based on the basic LAT protocol that evolves only one pair of SUs, we extended the size of secondary network in Chap. 3. As the first step, we considered the cooperative spectrum sensing under the LAT protocol, in which the interference among cooperative SUs becomes the unique feature in this scenario. A feasible cooperation scheme has been provided and proved effective in Sect. 3.1. Then, we

© The Author(s) 2016
Y. Liao et al., *Listen and Talk*, SpringerBriefs in Electrical and Computer
Engineering, DOI 10.1007/978-3-319-33979-5_5

considered more complicated networks with multiple contending users. Both distributed and centralized spectrum sharing and resource allocation mechanisms have been presented, and analyzed in Sect. 3.2.

Apart from the FD cognitive networks, which is a representative in vertical spectrum sharing, we also tried to apply the idea of LAT into horizontal spectrum sharing, featured by the WiFi networks. In Chap. 4, we studied the cross-layer protocol design for FD WiFi networks based on the basic idea of simultaneous sensing and transmission. Imperfect sensing brought by self-interference in PHY layer has been considered in the design, leading to adaptation or compensation in MAC layer. We showed by both analytical and simulation results that with the help of FD, the average collision length among contending users can be significantly reduced compared with the contentional CSMA/CA scheme, and the spectrum utilization efficiency can be improved a lot.

5.2 Research Challenges and Future Works

In fact, regardless of all the promising features that the FD CRNs hold, there still exist some challenges in the design and implementation of them. In this section, we briefly summarize the main research problems for FD-CR communication systems as well as the possible solutions. Also, some of the inspiring future research topics are listed at the end. Similar to the traditional wireless systems, multi-dimensional resources on space, time, frequency, and power need to be properly managed to optimize the overall system performance. Specially, FD communication provides another dimension of resource and its performance is also greatly affected by the RSI.

5.2.1 Signal Processing Techniques

5.2.1.1 Spectrum Sensing

The non-cooperative narrowband sensing has been elaborated in Chap. 2. However, the degradation of sensing performance becomes unbearable when transmit power is high. Cooperative sensing, as analyzed in Sect. 3.1 is one of the promising solutions for this problem. However, when employing cooperative spectrum sensing in FD-CRNs, there is interference from the transmitting SU to other SUs, which degrades the local sensing performance of the cooperative SUs to some degree, and the selection of the cooperative nodes may be different from the conventional HD scenario, i.e., the closest nodes may not be proper choices for cooperation due to the strong interference, and some further nodes may be better.

Also, since CRNs will eventually be required to exploit spectrum opportunities over a wide frequency range, a FD-enabled *wideband sensing scheme* is needed.

With the impact of RSI, the original sparsity, which is the base for conventional wideband sensing scheme, would change, and the whole sensing scheme may be different.

5.2.1.2 Multi-Antenna Techniques

If multiple antennas are equipped at the FD-CRs, beamforming and antenna selection can be employed to further improve the secondary network performance:

- Transmit beamforming: transmit beamforming is used to control the directionality of transmission in order to provide a large antenna array gain in the desired directions. For a FD cognitive MIMO system, the transmit antenna set at each FD-CR node can perform transmit beamforming to simultaneously transmit information and reduce the interference to its own received sensing signals. The design is to jointly optimize the sum rate of the system. If the FD cognitive AP node, which serves a group of users, is equipped with multiple antennas, it may be able to support multiple downlink transmissions while maintaining reliable sensing performance by using certain structure of antennas to minimize the RSI.
- Antenna selection: for a FD-CR node, especially a node with multiple antennas, each antenna can be configured to sense (receive) or transmit the signals. This creates an important problem to optimally select the antenna configurations optimize the system performance [1]. In a FD cognitive MIMO system, the problem is to choose one group of antennas to sense the spectrum, one group to transmit and the rest to receive signals from another SU simultaneously. Such a combinatorial problem becomes much complicated as the number of antennas increases. Similarly, in a general FD cognitive relay system, each relay can adaptively select its sense and transmit and receive antennas based on the instantaneous channel conditions to achieve reliable sensing as well as maximum SINR in transmissions.

5.2.2 Dynamic Spectrum Access and Resource Management

When multiple FD-SUs coexist in a network, how to detect and share the temporal and spectral resources to reduce collision or waste of spectrum holes and mitigate interference to the primary network becomes an urgent problem. DSA mechanism is one of the key techniques in the CRNs, through which, cognitive wireless nodes are able to adaptively and dynamically transmit and receive data in a changing radio environment harmoniously.

In Sect. 3.2, feasible distributed and centralized DSA schemes have been proposed and analyzed for FD-CRNs. However, these mechanisms can still be further optimized. Also, there are many other problems remaining uninvestigated. For example, when the primary network operates on multiple orthogonal channels, how to perform channel selection in a distributed way, and how to design a proper rendezvous mechanism so as to fully explore the advantage brought by the FD technique become quite important.

Apart from the designs of DSA schemes and channel allocation methods, power control, which is commonly deployed in traditional multi-user communication systems to optimize system performance such as link data rate, network capacity and coverage, also rises as a key challenge in FD CRNs. Unlike traditional wireless networks, FD-CR communication suffers from the RSI which deteriorate sensing performance when transmit power increases. Therefore, the corresponding power control algorithm needs to be properly redesigned in order to optimize system performance. Different power constraints, e.g. *total* or *individual* transmit power, will lead to different designs and final solutions. Moreover, as detailed below, different FD-CR systems require different power control algorithms:

- FD cognitive MIMO system: the antennas at the FD node are divided into sensing, receive and transmit antenna sets with individual power constraints. Considering the RSI at the sensing set, the optimal power pouring mechanism can be significantly different from the conventional water-filling.
- FD cognitive relay system: the relay is under individual power constraint, and both the relayed signals and sensing results are corrupted by the RSI. Increasing the transmit power at the relay will increase the SNR at the destination, but on the other side decrease the accuracy of sensing and blur the received signals from the source. The optimization needs to consider these issues.

5.2.3   Coexistence of Multiple Systems

Spectrum sharing has been recognized as a key remedy for the spectrum scarcity problem, especially after the successful deployment of WLAN and WPAN devices on an unlicensed band (e.g. ISM band). However, severe performance degradation has been observed when heterogeneous devices share the same frequency band due to mutual interference rooted in the lack of coordination. The cooperative busy tone (CBT) algorithm allows a separate node to schedule a busy tone concurrently with the desired transmission, and thereby improving the visibility among difference sorts of devices [2]. But preventing the busy signal from interfering the data packet still remains a problem. By deploying the FD techniques, the coexistence between heterogeneous networks may become more flexible. The research problem is to further reduce the RSI impact and realize efficient spectrum access management.

5.3 Applications and Future Research Topics

5.3.1 FD MIMO Networks

As shown in Fig. 5.1, it consists of a pair of FD MIMO transceivers, nodes A and B, where each node is equipped with multiple antennas (N), respectively. In each node, some antennas (N_s) are used for sensing, some (N_t) for data transmission, and some (N_r) can be used for receiving data from the other CR node. Both nodes operate in the same frequency band at the same time. Hence, if $N_s = N$, the system becomes the traditional CR with LBT; When all these three parameters are employed, this system supports bi-directional communication while sensing, but the interference is quite complicated among the antennas.

5.3.2 Cooperative Networks

Cooperation among different communication nodes is regarded as a promising method of enhancing the robustness of the communication system, mitigating interference among communication links, and improving quality of service in general. When several FD SUs cooperate with each other for sensing and joint scheduling, the sensing accuracy can be significantly improved, which has been roughly discussed in Sect. 3.1. The key difference of cooperation among FD SUs from the conventional cooperation in HD networks is the complicated interference brought by simultaneous sensing and transmission. As shown in Sect. 3.1, the interference from the transmitting SU to other cooperative SUs leads to severe degradation of sensing performance. Thus, how to select cooperative SUs, and

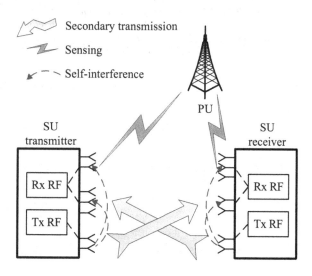

Fig. 5.1 Full-duplex cognitive MIMO system

how to combine the local sensing reports from different SUs regarding their location information are of great importance, and some new mechanisms need to be proposed. For example, for cooperative sensing, SUs that are too close to the transmitting SU may choose not to report their local sensing results, since their sensing is severely interfered. This kind of consideration leads to new design of cooperation schemes, which can be significantly different from conventional cooperations.

5.3.3   Heterogeneous Networks

Spectrum sharing has been recognized as a key remedy for the spectrum scarcity problem, especially after the successful deployment of WLAN and WPAN devices on an unlicensed band (e.g. ISM band). However, severe performance degradation has been observed when heterogeneous devices share the same frequency band due to mutual interference rooted in the lack of coordination. The cooperative busy tone (CBT) algorithm allows a separate node to schedule a busy tone concurrently with the desired transmission, and thereby improving the visibility among difference sorts of devices [2]. But preventing the busy signal from interfering the data packet still remains a problem. By deploying the FD techniques, the coexistence between heterogeneous networks may become more flexible. The research problem is to further reduce the RSI impact and realize efficient spectrum access management.

References

1. M. Zhou, H. Cui, L. Song, and B. Jiao, "Transmit-Receive Antenna Pair Selection in Full Duplex Systems," *IEEE Wireless Communications Letters*, vol. 3, no. 1, pp. 34–37, Feb. 2014.
2. X. Zhang and K. G. Shin, "Enabling Coexistence of Heterogeneous Wireless Systems: Case for ZigBee and WiFi," in *Proc. ACM International Symposium on Mobile Ad Hoc Networking and Computing (MobiHoc)*, New York, US, May 2011.

Printed in the United States
By Bookmasters